Texts in Philosophy
Volume 12

Philosophical Perspectives on Mathematical Practice

Volume 3
Monsters and Philosophy
Charles T. Wolfe, ed.

Volume 4
Computing, Philosophy and Cognition
Lorenzo Magnani and Riccardo Dossena, eds.

Volume 5
Causality and Probability in the Sciences
Federica Russo and Jon Williamson, eds.

Volume 6
A Realist Philosophy of Mathematics
Gianluigi Oliveri

Volume 7
Hugh MacColl: An Overview of his Logical Work with Anthology
Shahid Rahman and Juan Redmond

Volume 8
Bruno di Finetti: Radical Probabilist
Maria Carla Galavotti, ed.

Volume 9
Language, Knowledge, and Metaphysics. Proceedings of the First SIFA Graduate Conference
Massimiliano Carrara and Vittorio Morato eds.

Volume 10
The Socratic Tradition. Questioning as Philosophy and as Method
Matti Sintonen, ed.

Volume 11
PhiMSAMP. Philosophy of Mathematics: Sociological Aspects and Mathematical Practice
Benedikt Löwe and Thomas Müller, eds.

Volume 12
Philosophical Perspectives on Mathematical Practice
Bart Van Kerkhove, Jonas De Vuyst and Jean Paul Van Bendegem, eds.

Texts in Philosophy Series Editors
Vincent F. Hendriks   vincent@hum.ku.dk
John Symons   jsymons@utep.edu
Dov Gabbay   dov.gabbay@kcl.ac.uk

Philosophical Perspectives on Mathematical Practice

edited by

Bart Van Kerkhove

Jonas De Vuyst

and

Jean Paul Van Bendegem

© Individual author and College Publications 2010. All rights reserved.

ISBN 978-1-904987-59-8

College Publications
Scientific Director: Dov Gabbay
Managing Director: Jane Spurr
Department of Computer Science
King's College London, Strand, London WC2R 2LS, UK

http://www.collegepublications.co.uk

Original cover design by orchid creative   www.orchidcreative.co.uk
Printed by Lightning Source, Milton Keynes, UK

All rights reserved. No part of this publication may be reproduced, stored in a retrieval system or transmitted in any form, or by any means, electronic, mechanical, photocopying, recording or otherwise without prior permission, in writing, from the publisher.

Contents

Editors' Preface vii

1   The Unity of Mathematics: Distinctive Characteristic or Exaggerated Claim?
John W. Dawson, Jr. 1

2   Construction, Articulation, and Explanation: Phases in the Growth of Mathematics
Madeline Muntersbjorn 19

3   Laplace, Lakatos, and Luck: Accommodating Extrinsic Justification in Quasi-Empiricism
Ian Dove 43

4   Mathematical Concepts in Computer Proofs
Koen Vervloesem 61

5   For a Philosophy of Mathematical Practice
Gianluigi Oliveri 89

6   *In* or *About* Mathematics? Concerning Some Versions of Mathematical Naturalism
    José Ferreirós 117

7   Pragmaticism as an Anti-Foundationalist Philosophy of Mathematics
    Ahti-Veikko Pietarinen 155

8   Mathematical Knowledge as a Case Study in Empirical Philosophy of Mathematics
    Benedikt Löwe, Thomas Müller, and Eva Müller-Hill 185

9   Who Speaks Mathematics? A Semiotic Case Study
    Roy Wagner 205

10 Diagrammatic Reasoning in Euclid's *Elements*
    Danielle Macbeth 235

11 Observations on Sick Mathematics
    Andrew Aberdein 269

12 Two Types of Mathematization
    Johannes Lenhard and Michael Otte 301

13 Numerical Analysis and Its (Invisible?) Role in Mathematical Application
    Anthony F. Peressini 331

About the Authors 351

Editors' Preface

Bart Van Kerkhove
Jonas De Vuyst
Jean Paul Van Bendegem

This is the last of two volumes deriving from the second *Perspectives on Mathematical Practices* conference (PMP2007), which was held at the Vrije Universiteit Brussel (VUB), Belgium, from 26 to 28 March, 2007. Whereas a first volume (Van Kerkhove [2009]) has already addressed various aspects of the question how the philosophy of mathematics relates to the history of mathematics, the present volume rather focuses on philosophical matters themselves. The papers have been organised in four clusters, which we have however chosen not to formally separate in distinct parts of the book. The reason is that many, if not most, of the contributions simply resist being exclusively classified in either of the clusters distinguished by us. Before all, we have sought a way to present the available material to the reader in an orderly manner, the result being just one of the possible ways of achieving this.

The cluster of four papers opening this volume touches upon the complexity of mathematics as a scientific phenomenon. From an explicitly descriptive point of view, a number of quite familiar topics (at least to the philosopher of mathematical practice) are addressed here. First of these is the perceived unity of the mathematical sciences. John Dawson's findings, in Essay 1, are clear: In whatever sense one intends to take it (e.g., as methodological transferability, logical consonance, conceptional stability, or intersubjective agreement), and striking as it might be at the surface, by and large, mathematical unity is an illusion that upon closer inspection evaporates. Madeline Muntersbjorn, in Essay 2, favours another, more pluralistic approach to this issue, by taking mathematical unity as organic and dynamic in nature. On her view, mathematical kinds are robust but mutable, leaving room for a variety of philosophical positions dealing with the continuity inherent to the growth of mathematical knowledge. Muntersbjorn herself goes on to propose a possible model for the type of cyclic changes in mathematical practice, comprising phases of construction, articulation and explanation respectively. In their respective contributions, Ian Dove and Koen Vervloesem take us "beyond" Lakatos, by pointing out the importance of bottom-up truth-flow in mathematics. It is indeed common knowledge that Lakatos defended the upward flow of falsity in mathematics, i.e. quasi-empiricism, inspired by Popper's falsificationism. Less well known, however, is his case for induction, or rather abduction, i.e. the upward flow of truth in mathematics. Dove, in Essay 3, makes the argument for the general case, taking Lakatos as an evident initial guide, but also depending on Penelope Maddy, with an application to Laplace's work on divergent series. Vervloesem, in Essay 4, has another look at computer proofs, with a special eye on their distinctive role, as experimental tools, in the conceptual dynamics of mathematics. The analysis provides new insights into the limitations of computers as theorem provers.

The papers gathered in the informal second part of the book all probe for the philosophy (or philosophies) accompanying the acknowledgement of mathematical practice. Essay 5, by Gianluigi Oliveri, has been conceived as a true manifesto "for a philosophy of mathematical practice." It distinguishes two traditions or basic approaches in the philosophy of mathematics, an analytic and an historical one. Today, within both traditions, movements "back" to mathematical practice can be identified. Furthermore, particularly since the work of the aforementioned Imre Lakatos, the very separation of both traditions is being increasingly challenged, philosophy being more than abstract speculation, and history more than impartial descriptive reconstruction. Oliveri additionally contrasts internal and external histories of mathematical developments, the first being largely normative in doing justice to mathematical content, the second also appealing to "psycho-socio-cultural" context to that effect. On the basis of all these observations, the author distills a number of principles for the study of mathematical practices. José Ferreirós, in Essay 6, finds that, as part of a more general vogue in the past few decades, and thus having been used to denote a variety of incompatible philosophical positions, the label 'mathematical naturalism' has lost most of its distinctive meaning. Ferreirós instead proposes to make our way back to the basics of the doctrine, which comprises of non-foundationalism and an appeal to the natural sciences, most notably biology, to characterize knowledge. This, he says, would exclude a number of very interesting but also confusing positions, such as Maddy's or Bloor's, as worthy of the label in question. This, again, without therefore excluding the importance of the methodological or social factors they bring into focus, although Ferreirós does warn against cheap reductionism. As for Quine's and Kitcher's versions of naturalized epistemology, Ferreirós prefers the former's hypothetico-theoretic to the latter's empirico-idealistic character of the basic mathematical building blocks, viz. sets. He goes on to take us beyond Kitcher's empiricism *cum* Kuhnianism, making a case for the interconnectedness of coexisting practices rather than any philosophical assumptions as being the right guides for theoretical developments.

In the next two essays, completing the second cluster, two more specific and quite original proposals for a (partial) philosophy of mathematical practice are made. First, in Essay 7, Ahti-Veikko Pietarinen, on the basis of published and unpublished work of Charles Sanders Peirce, sketches a pragmaticist account of mathematics, which is utterly diagrammatic and iconic in nature, and is shown to stand in stark contrast to the most traditional of foundationalist philosophies but also to other, more recent proposals in the field, including structuralism, fictionalism and quasi-empiricism. And in Essay 8, Löwe, Müller, and Müller-Hill further explore an empirical philosophy of mathematics as conceived by the first two authors. The general idea behind this project is that, as the adequacy of any philosophical theories is not at all independent of the phenomena they refer to, proper philosophy of mathematics should solicit empirical, e.g. sociological or psychological, work adding to its strength. In the present paper, the focus is on mathematical knowledge, by developing a case study, on the basis of a questionnaire, about the role of formal proofs in mathematical practice.

Moving to a third cluster of contributions, one may find a number of perspectives on mathematical proof considered primarily as a text. The point of departure of Essay 9, by Roy Wagner, is precisely the perceived lack of the mathematical text as a given in existing (even semiotical) explorations of the role of mathematical subjects. In a first attempt to rectify this, he sets out an analysis of the enunciative position in Gödel's famous (first) incompleteness theorem, mainly drawing on Barthes and Foucault, and leading to the remarkable conclusion that the particular text under consideration makes sense thanks to a variety of conflicting voices and modalities. In the next contribution, Essay 10, Danielle Macbeth explains why, despite a shift in mathematical production since early modernity from diagrammatic demonstration to computation and formal symbolism, Euclidean geometry, the paradigm of ancient practice, is by no means flawed. It is instead a robust mode of mathematical inquiry, Macbeth argues, rooted in a system of natural deduction (instead of axiomatics) that allows one to actualize a vast array of general (instead of particular) conclusions about geometrical concepts. Concluding this part of the book, Andrew Aberdein, in Essay 11, discusses the argumentative structure of mathematical texts, more particularly the standard classification of correct versus erroneous arguments, due to E.A. Maxwell, which harbours mathematical fallacies as false results deriving from (seemingly) sound methods, and howlers as true results (luckily) deriving from unsound methods. Following his discussion, Aberdein proposes a richer and subtler typology of mathematical error, applying argumentation schemes, such as those from "verbal classification" or "popular opinion."

Concluding this proceedings volume, then, in a fourth and final cluster, are two papers focusing on practices of mathematical application. In their Essay 12, Johannes Lenhard and Michael Otte distinguish two fundamental types of "mathematization," concording with the dichotomy, identified by Hintikka, between two conceptions of Leibniz' logical project. In a first interpretation, the search for certainty develops along the lines of logical principles, language being a reflection of our conceptual structures, whereas in a second interpretation, certainty is sought after in human construction and measurement, language being a method mirroring our actual reasoning processes. This dichotomy is illustrated by three case-studies from the history of applied mathematics, opposing Leibniz to Newton concerning natural philosophy, Grassmann to Ampère concerning electricity theory, and Wiener to von Neumann concerning computer theory. Closing the book, in Essay 13, Anthony Peressini discusses the nature and importance of numerical analysis. He points out that almost anything that happens in mathematics today, when going beyond mere counting, involves some of its techniques, albeit often on a hidden level, largely because of the use of computing devices of all sorts. This, Peressini argues, also pertains to applied mathematics, where—visible or not—it performs either an instrumental, essential or exploratory role, depending on the availability of exact solutions.

All this material having been presented, where do we stand in general, and moreover, where do we go from here? It has been observed many times before that, as yet, there are no encompassing, integrated theories of mathematical practice on the market. However, with a rising supply in recent years of collections of studies including the present one and its PMP-predecessors (Van Kerkhove and Bendegem [2002] and [2007]), by now we at least have available the germs of a number of competing approaches. Paolo Mancosu ([2008]) has recently distinguished two important such approaches (not aiming for completeness): "maverick" anti-foundationalism in the wake of Imre Lakatos, and epistemic naturalism in the wake of W.V.O. Quine, tendencies which are today most forcefully represented by David Corfield (see his [2003]) and Penelope Maddy (compare Ferreirós' comments referred to above) respectively. Mancosu however situates the most interesting approaches in between these rather extreme tendencies, citing his own as well as others' collections of papers (Mancosu [2008]; Ferreirós and Gray [2006]). As he argues, the mavericks were absolutely right to accuse the foundationalists of remaining largely irrelevant to philosophy and history of mathematics. That is because they straightforwardly neglected important topics pertaining to mathematical knowledge, most notably topics with a bearing upon its "dynamics," such as progress or explanation, while upholding the eternal and non-fallible nature of mathematics. As for the observation that, despite its justified criticism, the maverick tradition as a whole has not (yet) managed fundamentally to redirect philosophy of mathematics, Mancosu's answer is twofold. First, as for their "destructive" work, the mavericks have generally been too polemical towards both foundationalists and traditional analytic philosophers of mathematics. Second, as for their "constructive" work, their scope and depth have remained too restricted, and the metaphilosophical ambitions disproportionately big. (The latter, not the former, criticism would also apply to Maddy.)

This can be turned into a more general observation concerning studies of mathematical practices. For to witness, as we currently do, a variety of schools elaborating their philosophical frameworks, and trying to sort out their differences in the course of doing so, is also to be constantly reminded of the fact that a lot of aspects, extremely relevant to this task, remain dramatically underexamined. It shows a sense of proportion, then, to acknowledge that a couple of decades after the appearance of the famous case-study *Proofs and Refutations* (Lakatos [1976]), the primeval book in mathematical practice studies, it is in a way still early days. Mancosu's proposed remedy is straightforward, and implicit to his as well as to the present collection: At least for now, it is not complicated metaphilosophical claims that should be at stake, but rather, more modestly, epistemic (often methodological) issues arising naturally from within mathematical practice. This results in a more "conciliatory" position towards both mathematical foundationalism and analyticism, which conserves energy required to cover a greater number of issues in sufficient detail. One may hope that the newly established *Association for the Philosophy of Mathematical Practice* (APMP), an initiative of, among others, Mancosu, Ferreirós, Gray, and the third editor of this volume, in years to come will substantially contribute to facilitating this.

Finally, a few words of appreciation. For a proceedings volume does not come about because of the efforts of the editors alone. To begin with, we are extremely grateful for all institutional and personal contributions to what in our eyes was a very successful PMP2007 conference. This includes the generous sponsors of the event: Research Foundation–Flanders, Brussels Capital-Region, National Centre for Research in Logic–Belgium, as well as its (co-)organizers Belgian Society for Logic and Philosophy of Science, Wissenschaftliches Netzwerk PhiMSAMP, and Centre for Logic and Philosophy of Science at the Vrije Universiteit Brussel. Further, we were very fortunate to leave local organization largely in the safe hands of our precious colleagues Patrick Allo, Ronny Desmet, and Karen François. And as there is simply no event whatsoever without an interested and interesting audience, let us hereby also thank all participants to PMP2007 (including the authors of this book). We sincerely hope that this conference series, started in 2002 with the initial PMP, may live on, possibly at other locations.

Bibliography

Corfield, D. [2003]. *Towards a Philosophy of Real Mathematics*. Cambridge: Cambridge University Press.

Ferreirós, J. and J.J. Gray, eds. [2006]. *The Architecture of Modern Mathematics*. Oxford: Oxford University Press.

Lakatos, I. [1976]. *Proofs and Refutations. The Logic of Mathematical Discovery*. Cambridge: Cambridge University Press.

Mancosu, P., ed. [2008]. *The Philosophy of Mathematical Practice*. Oxford: Oxford University Press.

Van Kerkhove, B., ed. [2009]. *New Perspectives on Mathematical Practices. Essays in Philosophy and History of Mathematics*. Singapore: World Scientific.

Van Kerkhove, B. and J.P. Van Bendegem, eds. [2002]. *Logique et Analyse* **45** (179–80). *A Selection of Papers Presented at the Perspectives on Mathematical Practices Conference 2002*. Appeared in 2004.

— eds. [2007]. *Perspectives on Mathematical Practices. Bringing Together Philosophy of Mathematics, Sociology of Mathematics, and Mathematics Education*. Vol. 5. Logic, Epistemology, and the Unity of Science. Dordrecht: Springer.

ESSAY 1

The Unity of Mathematics
Distinctive Characteristic or Exaggerated Claim?

John W. Dawson, Jr.

Many commentators have cited the unity of mathematics as one of its most notable features. Some have claimed such unity to be a distinctive characteristic of mathematics, in contrast, e.g., to the natural sciences. And Michael Atiyah has even asserted that "the ultimate justification for doing mathematics is intimately related with its overall unity" (Atiyah [1985], quoted in Gowers [2000]: 75). But what is *meant* by unity has not always been made clear, and those who have clarified their meaning have not always had the *same* conception in mind. In addition, different observers have disagreed on whether the phenomena that have been adduced as manifestations of a particular conception of unity are in any sense remarkable.

In what follows we consider various ways in which mathematics has been deemed to exhibit unity. We conclude that the belief that mathematics is distinguished from other natural sciences by its unity is largely an illusion.

1 Manifestations of Unity

Issues concerning unity and unification in the sciences, and the relation of unification to explanation, have received a good deal of attention from philosophers of science (see, for example, Morrison [2000]). More recently, a growing literature on explanation and unification within mathematics has been produced by philosophers of mathematics. Here, however, I shall focus primarily on statements by mathematical practitioners.

One sense in which mathematics exhibits unity was articulated by Peter Lax, who asserted:

> There is no more convincing proof of the unity of mathematics than notions developed in one field turning out to be the key in solving outstanding problems of another seemingly unrelated field. (Lax [1986])

I shall refer to this aspect of unity as the *transferability of notions and methods* among different mathematical disciplines. In Lax's opinion, such transferability "is especially stunning when notions of applied mathematics are used to answer questions of very pure mathematics"[1] (ibid.)—a view in accord with that of the applied mathematician R.W. Hamming, who contended that "The enormous usefulness of the same pieces of mathematics in widely different situations has no rational explanation (as yet)" (Hamming [1980]). And more recently, in a pregnant discussion of the role of analogy in mathematics, the philosopher David Corfield has described the "sometimes mysterious, yet common, use of concepts drawn from one area to illuminate another body of theory" as "a riddle begging philosophical treatment" (Corfield [2003]: 81).

E.H. Moore, on the other hand, thought that "The existence of analogies between the central features of various theories" was not at all mys-

---

[1] The Dirac δ-"function" is one notable example.

terious, but rather constituted evidence of "the existence of a general theory which underlies the particular theories and unifies them with respect to their central features" (Moore [1910]). And Nicolas Bourbaki, somewhat patronizingly, declared that

> Whereas the superficial observer sees . . . several quite distinct theories lending one another 'unexpected support' through the intervention of a mathematician of genius, the axiomatic method teaches us to look for the deeper-lying reasons [why that is so]. (Bourbaki [1950])

A second way in which mathematics has been deemed to exhibit unity is what I shall call *consonance of results*: the fact that different theories, developed at different times for different purposes and employing different concepts and methodologies, seem to agree in their consequences, apart from terminological differences, wherever their subject matters overlap—a sort of 'analytic continuation' across different mathematical disciplines. As Max Dehn put it, "Most [mathematical] results are so involved in the general web of theorems, they can be reached in so many ways, that their incorrectness is simply unthinkable" (Dehn [1983]).

This aspect of unity, which stresses how *different* methods so often lead to the *same* conclusions, is distinct from the *transferability* of notions stressed above. For though both transferability and consonance involve the application of methods outside the domain in which they were originally developed, transferability *per se* need not involve rederivation of known results. The use of algebraic methods to establish the impossibility of certain constructions with straightedge and compass, for example, settled some long-standing problems that had *resisted* geometric solution. Consonance, on the other hand, implies *lack of conflict* with results already obtained by other means. For example, the Fundamental Theorem of Algebra can be derived from many different perspectives (cf. Fine and Rosenberger [1997]), including the theory of complex contour integration (Liouville's theorem) and Galois theory, both of which were developed (for very different purposes) well after Gauss' original proof of that theorem (1799). What is striking about such consonance, as Hilary Putnam ([1967]) has remarked, is "the number of ways of expressing what is in some sense the same fact . . . while apparently *not* talking about the same objects" (my emphasis).[2]

Judson Webb has pointed out that such consonance of results was a matter of particular concern to David Hilbert:

> [I]n his work on radiation theory and elementary optics . . . consistency proofs first take on for Hilbert the significance that they characteristically have in his later work, namely as showing not only that some theory is consistent in and of itself, but that *it does not contradict the laws of some more elementary 'neighboring' theory to which it can be applied.*
> (Webb [1980]: 139–40, his emphasis)

A third claimed aspect of mathematical unity is the *stability of mathematical concepts over time*, in contrast to the conceptual instability so prominent in the history of the natural sciences. The conceptual bases of physics, chemistry, and biology, for example, have changed so radically over the centuries that physicists, chemists, and biologists presumably cannot read past writings in their fields with the ease or understanding that mathematicians can. To some extent, however, that presumption is an illusion, resulting from the uncritical reading of latter-day concepts *into* earlier mathematical writings.

Infinitesimals are a case in point. Like such entities as phlogiston in chemistry and the luminiferous aether in physics, infinitesimals were once a central conception in mathematics. In the wake of Berkeley's withering characterization of them as "ghosts of departed quantities," they were eventually discarded. Later, however—unlike phlogiston and the aether—the notion of infinitesmals was resurrected and restored to good standing through Abraham Robinson's creation of non-standard analysis. Nevertheless, as Henk Bos has argued in detail in the appendix to his [1974], it is wrong-headed to claim that Leibniz had non-standard numbers in mind when he spoke of infinitesimals (or that Cauchy did

---

[2]Recognition that two seemingly disparate statements *do* in some sense express "the same fact" is in itself a quintessential mathematical act that often requires considerable sophistication. Modern abstract formulations of classical mathematical results are well-known exemplars.

when he asserted that the limit of a convergent sequence of continuous functions is necessarily continuous). Rather, the use of non-standard analysis (based on the latter-day concept of ultrapowers) to legitimize arguments involving infinitesimals is another example of the *consonance* of results obtained by different means.

In addition (as one commentator has perceptively noted), stability in and of itself need not imply unity at all. Indeed, *rigid* stability of mathematical notions might well *obstruct* unification, by preventing concepts from being sufficiently flexible to permit common formulations. (Cf. the remarks in section 2 below concerning the invariance of theorems under generalization.)

Other aspects of mathematical practice that have been adduced as manifestations of mathematical unity are the extent of intersubjective agreement among mathematicians' intuitions, the extent of intercommunication among specialists in different subfields, and, especially, the expressibility of all of mathematics within the language of set theory. With regard to the first of those aspects, Hao Wang went so far as to claim that such intersubjective agreement is "the fundamental empirical datum for the formulation and the evaluation of every account of the nature of mathematics" (Wang [1996]: 210); and in support of the second is the fact that there are no gatherings in physics, biology, or chemistry that are comparable in universality to the quadrennial international congresses of mathematicians. But contrary to Wang, it is not clear that the commonality of intuitions among mathematicians differs essentially from that observable among practitioners in other fields; and since most mathematicians have undergone years of rigorous training in the same core mathematical subjects and techniques, one may question whether their subjective agreement is anything more than a reflection of their having absorbed a common body of knowledge and perspectives from their teachers. (Indeed, individuals such as L.E.J. Brouwer, whose mathematical conceptions differ radically from those of the mathematical community, are likely to be misunderstood and perhaps even to be written off as cranks. Brouwer himself might well have been had he not established important results in topology prior to his espousal of intuitionism.) As for the solidarity exemplified at the international congresses, apart from the ceremonies at which the Fields Medals are awarded and a few plenary lectures by the most eminent contemporary mathematicians, most participants attend only presentations that are closely related to their own fields of specialization.[3]

The possibility of expressing all of contemporary mathematics within the framework of axiomatic set theory is, to be sure, evidence of some underlying unity. On closer scrutiny, however, several objections may be made. First, one must distinguish possibility from actuality. Second, as Yehuda Rav ([1999]: n. 20) has stressed, to acknowledge that "current mathematical theories can be *expressed* in first-order set-theoretical language" does not compel belief that all "current *conceptual proofs*"—the very core of mathematical endeavor, in his view—"can be *formalized as derivations*" within that framework (his emphasis).[4] Most importantly, if the framework for set theory is taken to be first-order ZFC, as it usually is, then expressibility within that framework may be artificial to the point of distortion. For example, the existence of non-standard models of first-order arithmetic conflicts with Peano's informal (second-order) proof that his axioms uniquely characterize the natural numbers. So if number theory is formalized within first-order ZFC, that theorem must simply be discarded. Nor is that an isolated example: Stewart Shapiro ([1991]) has argued persuasively that "Second-order logic provides better models of important aspects of mathematics, both now and in recent history, than first-order logic does." But as a vehicle for expressing all of mathematics, second-order logic is itself problematic, since it lacks many of the metatheoretical properties that logicians have come to cherish, such as completeness and compactness.[5]

---

[3]Though, as one referee remarked, "it's intriguing that those leading mathematicians giving the plenary lectures are [generally also] those whose work . . . performs the bridging between disciplines."

[4]For example, the eminent logician Jon Barwise noted that none of the formal models of proof so far proposed admit such "perfectly valid (and ubiquitous) form[s] of mathematical reasoning" as proofs by cases in which one case is established in detail and others are observed to follow by symmetry considerations. (Barwise [1989])

[5]More precisely, according to Lindström's Characterization Theorem, first-order logic is the strongest logic for which both the countable compactness theorem and the downward Löwenheim–Skolem theorem hold.

2 Some Caveats Regarding Transferability and Consonance

It seems indisputable that there are numerous examples of the transferability of concepts and methods among different areas of mathematical inquiry, and that there is consonance among the results obtained by different mathematical methodologies. But that such phenomena defy rational explanation, or that they constitute evidence of an underlying unity that distinctively characterizes mathematics, are claims that deserve more critical examination.

In many instances the transferability of notions exposes previously unexpected connections between seemingly unrelated fields, and so leads to a more unified perspective (and sometimes to major advances in those fields). But if mathematicians *believe* in the underlying unity of their subject, why should such transferability appear *surprising*, as it often does?

Transferability is certainly striking *when it succeeds*, but it is not likely to attract much notice when it fails. An unsuccessful attempt to carry out a proof with given means is, after all, a common occurrence that is not ordinarily of great significance; it is not like an *experimentum crucis* in the natural sciences. Rather, it is natural to attempt to apply techniques that have proved fruitful elsewhere in new contexts, and *sometimes* that attempt is successful. When it is not, new approaches are called for, whose development (which, contrary to Bourbaki, often *does* result from the "intervention of a mathematician of genius") may also lead to major advances. The question then becomes: Does success in extending known techniques occur disproportionately often, or is it a matter of skewed perspective?

To assert that instances of unification brought about by transferability of notions are evidence of some underlying *global* unity also raises the question of the *scale* on which transferability occurs. That certain basic principles of *logic*, for example, should be applicable in all areas of mathematics (and indeed, of science) hardly seems evidence of any distinctive or deep-seated mathematical unity. What is of interest are rather cases in which more specialized techniques or viewpoints, characteristic of and developed within one particular area of mathematics, have turned out to be useful, or even to unify, other areas of mathematics in which their applicability was unforeseen. Klein's *Erlanger Program*, for example, provided a group-theoretic perspective under which various different geometries could be harmoniously subsumed. Other substantive examples are given in Lautman [1937] (the significance of dimensional decompositions in the theory of functions, of non-Euclidean metrics in the theory of analytic functions, and of finite and discontinuous algebraic structures on the existence of functions of a continuous variable) and Corfield [2003] (the use of Hermitian metrics from differential geometry in number theory, of prime ideals in the spectrum of a ring of algebraic integers in knot theory, and of fields of algebraic numbers in the study of algebraic functions).

Lautman concludes (contrary to my own assertion in the preceding paragraph) that "the unity of mathematics is essentially that of the logical schemas which underlie its edifices."[6] But do such logical schemas really play a significant role in mathematical practice? They have certainly not done so throughout much of mathematical history. To be sure, in the last century and a half (the period from which all of Lautman's examples are taken) the search for general principles that underlie and unify particular theories (to paraphrase Moore's statement quoted earlier), and the belief that such principles are there to be found, has been a driving force. And there is no doubt that the conceptual abstraction and stress on axiomatic methodology resulting from that have led to some significant instances of unification. But many conceptual and methodological oppositions persist: finite versus infinite, discrete versus continuous, local versus global, algebraic versus analytic, commutative versus non-commutative, linear versus non-linear, topological versus metric, and geometric versus algebraic. And there are contemporary mathematicians of stature who do *not* believe that all such dichotomies are bridgeable.

A.R.D. Mathias, for example, has avowed that "there are *two* primitive mathematical intuitions, which might be called the geometrical and the arithmetical" (Mathias [1992*b*]; see also his [1992*a*]). While "cer-

---

[6]"[I]l résulte que l'unité des mathématiques est essentiellement celle des schémas logiques qui président à l'organisation de leurs édifices."

tain formal translations exist" between those two modes of thought, he contends that there are obstructions, such as the Banach–Tarski paradox, that prevent translation of the underlying intuitions. Likewise, Max Dehn thought that "in the case of very general conceptual constructs there arise *contradictions* between geometry . . . and the world of counting" (Dehn [1983], my emphasis). To what extent, then, does the purported unity of mathematics reflect the perspective of the beholder?

Bourbaki conceded that "the various attempts to integrate . . . mathematics into a coherent whole . . . have all been made in connection with . . . philosophical system[s], more or less wide in scope" that are based on "*a priori* views concerning the relations of mathematics with the twofold universe of the external world and the world of thought"; and since Bourbaki's position is itself philosophically grounded, it is open to the charge that it seeks to impose global coherence where none may in fact exist. Moreover, recognition of the importance of the axiomatic method in modern mathematical practice does not render a position such as Hamming's less tenable. Indeed, Hamming agreed that "one of the main strands of mathematics is the extension, the generalization, the abstraction . . . of well-known concepts to new situations." He noted, however, that "in the very process [of such generalization and abstraction] the definitions . . . are subtly altered, [so that] old proofs . . . may become false. . . . The miracle is that almost always the theorems are still true." (Hamming [1980])

In fact, the extent to which theorems exhibit invariance in the course of generalization or abstraction is not at all clear cut. In some cases, redefinition of terms causes the meaning of a theorem to change even though its statement remains unaltered. If done crassly, by restricting the scope of a term in an *ad hoc* fashion expressly to preserve the appearance of a theorem statement in the face of unforeseen counterexamples (what Imre Lakatos in his [1976] dubbed "the method of monster-barring"), there is nothing 'miraculous' about it. Alternatively, the conclusion of a theorem may be retained while new restrictive hypotheses are added (what Lakatos called "exception-barring"); but then, of course, the theorem is no longer the same.[7]

---

[7]In accord with Lakatos' views, Hamming remarked that "If you came into my

In other instances the meaning of terms is not *restricted* through redefinition, but *expanded*. Some theorems may thereby be generalized to new contexts (perhaps with new proofs), while their meanings (and the proofs previously given for them) remain unchanged in the original contexts. But broadening the meaning of a concept can also cause some earlier theorems to become *in*valid. The introduction of complex numbers, for example, affected the existence (and number) of solutions of polynomial equations. The modern concept of function, in which explicit defining formulas or mechanical constructions play no role, is another example. (The evolution of that conception, and the outcry against the 'pathological' functions it legitimized, is well known.) And the definition of polyhedron, analyzed in historical detail by Lakatos, is yet another.

The observed consonance among mathematical theories may at first seem harder to account for, since it appears to come about serendipitously. But once again, that appearance depends upon one's philosophical presuppositions.

Consider, for example, a recent, well-known example of consonance: the confluence of ideas concerning the notion of what constitutes an "effective" procedure. (See Gandy [1988] for a historical survey.) When Alonzo Church introduced the concept of λ-definability, he did not intend it to provide a characterization of all such procedures (that is, of the class of all recursive functions); indeed, as Stephen C. Kleene recalled in his [1981] memoir, it was not at all evident that even the *predecessor* function was λ-definable. On the other hand, the definitions of Herbrand/Gödel in terms of systems of equations, and of Turing in terms of finite-state machines mimicking the operations of a human computor, *were* intended to capture general notions of recursion or of computability, respectively, and it was in terms of the former that Church publicly announced his Thesis. But it was not obvious that all those definitions

---

office and showed me a proof that Cauchy's theorem was false I would be very interested, but I believe that in the final analysis we would alter the assumptions until the theorem was true" (ibid.). But Lakatos' theory seems unable to account for the transferability of notions between pure and applied mathematics, which Lax found so striking, because since applied mathematics is constrained by the real world, the extent of accommodation within it is strictly limited. See also the critiques of Lakatos' position by Feferman ([1978]) and Corfield ([2003]: 154–61).

characterized the *same* class of functions: The extensional agreement among the intensionally distinct notions came as a surprise to all concerned. To Gödel, for example (who held back from accepting Church's Thesis until after Turing's work), it was "a kind of miracle" that the definition of general recursive function was independent of the formalism chosen. He marveled, in particular, that "[T]he diagonal procedure d[id] not lead outside the defined notion" (Gödel [1946]). And Howard Aiken, one of the pioneers of computer science, remarked as late as 1956 that

> [S]hould [it] turn out that the basic logics of a machine designed for the numerical solution of differential equations coincide with the logics of a machine intended to make bills for a department store, I would regard th[at] as the most amazing coincidence I have ever encountered.
> (Quoted in Ceruzzi [1983]: 43 and Davis [1987]: 140)

It is worth noting, too, that the confluence of ideas in recursion theory did *not* arise either from internal organizational concerns or from a need to reconcile conflicting results—a circumstance that, from some philosophical perspectives, may make such consonance appear all the more mysterious. (How, for example, would a conventionalist like Carnap explain it?) From an empiricist or platonist perspective, however— according to which mathematics is ultimately grounded either in our experience of the material world or (as Gödel maintained) in our perception of a conceptual world apart from it—instances of consonance among different mathematical theories may be taken simply to reflect different *facets* of the same underlying material or conceptual reality. That we may at first be surprised by such consonance is then readily explained as well, since the same objects seen from different angles often appear very different. (And in perceiving the material or conceptual world around us we may only "see through a glass darkly.") But if so, why have some concepts of physics, which would seem to be *more* directly grounded in the world around us than those of mathematics, so far *resisted* unification?[8] (For further discussion of many issues arising

---

[8] One possible explanation was given by Eugene Wigner in his famous [1960] paper. He contended that the two foundations of modern physics, the theory of quantum phenomena and the theory of relativity, "have their roots in mutually exclusive

from comparison of mathematics and physics, see e.g. chapters 1 and 2 of Morrison [2000].)

From a formalist standpoint, such as that advocated by Bourbaki, the phenomenon of consonance raises the question whether suitable formal frameworks can be found that permit comparisons to be made among proofs based on methods from different subfields of mathematics. To date, however, apart from some conservative extension results, little progress has been made in that direction.

3 Perceptions of Unity or Disunity

Has the unity of mathematics always been manifest? Or has the perception of unity within mathematics changed over time?

It is easy to adduce examples of perceived *dis*unity within mathematics. To the ancient Greeks, for example, the discovery of incommensurability seemed to show incompatibility between arithmetic, the science of discrete quantities, and geometry, the science of continuous magnitudes. Indeed, to quote Max Dehn yet again, once the existence of irrational lengths is acknowledged, "Why *should* the rules for computing with whole numbers or fractions that admit of simple direct derivation [be expected to] hold for . . . 'inexpressible' ratios as well?" (Dehn [1983], my emphasis). What is truly remarkable is that they *do*, as Eudoxus was later able to show through his geometric theory of proportions. Theoretical unity of a sort was thus re-established, by subsuming arithmetic within geometry.

Later, in the wake of Fermat's *Isagoge* and Descartes' *Géométrie*, that perspective was reversed: Geometry underwent algebraization, which Mancosu in his [1996] has called "the most important event in seventeenth-century mathematics." In his *History of Analytic Geometry* (Boyer [1956]: 181), Carl Boyer notes that one advantage of the algebraic

---

groups of phenomena." With regard to the empiricist view of mathematics, Wigner thought that though it was "unquestionably true that the concepts of *elementary* mathematics . . . were formulated to describe entities which are directly suggested by the actual world," that did "not seem to be true of the more advanced concepts, *in particular the concepts which play such an important role in physics*" (my emphasis). Accordingly, it seemed to him "a miracle" that the "recklessness [with which] the great mathematician . . . exploits the domain of permissible reasoning and skirts the impermissible . . . does not lead . . . into a mass of contradictions." The consonance of mathematical results thus remained mysterious to him.

approach was that "many special cases [could] be included within one comprehensive formulation." But he stresses that the aim of both Fermat and Descartes was to use algebraic methods to facilitate geometric constructions rather than to discover new theorems in Euclidean geometry. For the latter, synthetic methods were considered both adequate and preferable (in part, no doubt, due to concern for *purity of method*—a concern in *opposition* to transferability of notions). The algebraization of geometry once again produced *theoretical* unification of a sort; but the underlying *methodological* opposition remained unresolved, and in the nineteenth century advances in projective and descriptive geometry came about through a reversion to synthetic techniques.

The history of non-Euclidean geometry provides a second instructive example. Euclid's fifth postulate was deemed less acceptable than his other postulates *not* because its truth was doubted, but because it appeared less evident, and so seemed less suitable to be taken as a basic principle. Accordingly, it was hoped that the fifth postulate would prove to be derivable from the others. Eventually, after centuries of effort had failed to yield such a derivation, Saccheri thought he had found an indirect proof: He (mistakenly) claimed to have derived a contradiction by replacing the fifth postulate by its negation.

When, in the wake of Saccheri's work, Beltrami and others described concrete surfaces whose geodesics failed to satisfy Euclid's fifth postulate, there was consternation—presumably because those models upset the belief that there could *be* only one geometry (a belief, that is, in unity). But how was that belief tenable, given that *spherical* geometry, whose geodesics fail to satisfy the parallel postulate, had long been in use for navigation?

According to Jeremy Gray in his [1989], what allowed spherical geometry to co-exist alongside Euclidean geometry without upsetting the notion that denial of the parallel postulate must lead to contradiction was that "spherical geometry treats of curved lines on curved surfaces and not of straight lines." That is, Euclidean and spherical geometries did not refer to the *same* entities, so there was *no perceived lack of consonance* between them.

More recently, in his [2007] article, David Stump has amplified Gray's conclusion. He points out that even *after* the construction of non-Euclidean models, rigorous consistency proofs of negations of the parallel postulate could not be given until a *formal conception* of geometry was adopted; not, that is, until the primitive terms of geometry came to be regarded as *uninterpreted* notions, relations among which were determined solely by the axioms. Stump traces the development of that perspective through the work of Plücker, Riemann, and Klein to its culmination in the work of Hilbert and Poincaré.

Historically, then, belief in the unity of geometry was shattered by the construction of non-Euclidean geometric models. Later, rigorous demonstrations of the formal consistency of non-Euclidean axioms became possible. What enabled such proofs was the development of a more general conception of what a geometry *is*; and with that change in perspective, Klein's *Erlanger Program* could be viewed as a restoration of unity.

These examples, among many that could be given, suggest that the perception of unity within mathematics has varied with time, and among different observers. Episodes of perceived disunity have repeatedly been succeeded by instances of unification. But methodological divisions remain (most notably, perhaps, between the methods of discrete, combinatorial mathematics, and those of analysis[9]); unification has, in some cases, only *partially* occurred (see Grosholz [2000] for a perceptive discussion of examples); and, as noted above, there are attitudes that are directly *opposed* to transferability of methods between subdisciplines.

4 Is Unity a Distinctive Characteristic of Mathematics?

Since the *pursuit* of unification is common to all the sciences, one must therefore ask: Does mathematics exhibit greater unity than other scientific fields, such as physics? In the introduction to the first edition (1928) of *The Theory of Groups and Quantum Mechanics* (translated as Weyl [1931]), Hermann Weyl thought there was "a plainly discernible parallelism between the more recent developments of mathematics and physics." In earlier centuries, he claimed, "occidental mathematics . . . had broken away from the Greek view" by regarding "the concept of number . . . as logically prior to the concepts of geometry." But, he

---

[9] Gowers ([2000]) addresses this "cultural divide" in detail.

continued, "the present trend in mathematics is clearly in the direction of a return to the Greek original; [for] we now look upon each branch of mathematics as determining its own characteristic domain of quantities." Similarly, "the continuum of real numbers has retained its ancient prerogative in physics for the expression of physical measurements . . . but the essence of the new . . . quantum mechanics is . . . that there is associated with each physical system a set of quantities" that is characteristic of that system.

Weyl seems to be saying (contrary to the interpretation propounded in Lautman [1937]) that modern mathematics and physics agree in their *lack* of unity. That seems at odds with the contrast often drawn between the *unity* of mathematics and the *dichotomies* so prominent within modern physics between wave and particle and between gravitational and nuclear forces. (Cf. Wigner's view quoted in Footnote 8 above). But the divide within mathematics between the realms of the continuous and the discrete seems comparable in many respects to that between wave and particle. And whereas the quest to unify the theories of the basic forces in nature has been and remains a central concern of contemporary physicists, bridging the gulf between the discrete and the continuous, or between the finite and the infinite, seems of little concern to many contemporary mathematicians.[10]

Are the connections between different mathematical fields (say, algebra and topology) closer than those between different areas of physics (such as mechanics, electrodynamics, and optics)? That seems hard to quantify, and a dubious contention. On the basis of the evidence presented herein, it seems more reasonable to conclude that the aspects of unity claimed for mathematics have been exaggerated and do not serve to distinguish the mathematical enterprise to any significant degree from other fields of scientific endeavor.

Acknowledgments

The ideas discussed in this paper were first presented at a symposium honoring the seventieth birthday of Professor Solomon Feferman, held

---

[10]On the other hand, Alain Connes contends that the "problem of treating continuous and discrete variables on the same footing is completely solved using the formalism of quantum mechanics." For details, see http://noncommutativegeometry.blogspot.com/2007/06/infinitesimal-variables.html.

at Stanford in December 1999. A later version was delivered to the conference Perspectives on Mathematical Practices, held in Brussels in March 2007. This revision has benefitted from comments of numerous readers and conference participants. I wish especially to thank Jeremy Avigad, Johan van Benthem, Martin Davis, Solomon Feferman, Emily Grosholz, Paolo Mancosu, Grigori Mints, Charles Parsons, Dana Scott, and two anonymous referees for constructive criticisms and suggestions.

Bibliography

Atiyah, Michael F. [1985]. "Identifying progress in mathematics." *The Identification of Progress in Learning*. Cambridge: Cambridge University Press. 21–41.

Barwise, J. [1989]. "Mathematical proofs of computer system correctness." *Notices of the American Mathematical Society*. Vol. 36. 844–51.

Bos, Henk J.M. [1974]. "Differentials, higher-order differentials and the derivative in the Leibnizian calculus." *Archive for the History of Exact Sciences* **14**. 1–90.

Bourbaki, N. [1950]. "The architecture of mathematics." *American Mathematical Monthly* **57**. 221–32.

Boyer, Carl B. [1956]. "History of Analytical Geometry." *Scripta Mathematica Studies*. New York: Scripta Mathematica.

Ceruzzi, P.E. [1983]. *Reckoners, the Prehistory of the Digital Computer, from Relays to the Stored Program Concept, 1933–1945*. Westport, CT: Greenwood Press.

Corfield, D. [2003]. *Towards a Philosophy of Real Mathematics*. Cambridge: Cambridge University Press.

Davis, M. [1987]. "Mathematical logic and the origin of modern computers." *Studies in the History of Mathematics*. Ed. by E.R. Philips. Vol. 26. Washington: Mathematical Association of America. 137–65.

Dehn, M. [1983]. "The mentality of the mathematician: A characterization." *The Mathematical Intelligencer* **5** (2). English translation of an address delivered 18 January 1928 to the faculty and students of the University at Frankfurt am Main. 18–26.

Feferman, S. [1978]. "The logic of mathematical discovery vs. the logical structure of mathematics." *PSA 1978*. Vol. 2. East Lansing, MI: Philosophy of Science Association. 309–27.

Fine, B. and G. Rosenberger [1997]. *The Fundamental Theorem of Algebra*. New York, Berlin, and Heidelburg: Springer Verlag.

Gandy, R. [1988]. "The confluence of ideas in 1936." *The Unviersal Turing Machine: A Half-Century Survey*. Ed. by R. Herken. Oxford: Oxford University Press. 55–111.

Gödel, K. [1946]. "Remarks before the Princeton Bicentennial Conference on problems of mathematics." *The Undecidable*. Ed. by M. Davis. New York: Hewlett and Raven Press. 84–8.

Gowers, W.T. [2000]. "The two cultures of mathematics." *Mathematics: Frontiers and Perspectives*. Ed. by V. Arnold et al. Providence: American Mathematical Society. 65–78.

Gray, J. [1989]. *Ideas of Space, Euclidean, Non-Euclidean and Relativistic*. 2nd ed. Oxford: Oxford University Press.

Grosholz, E. [2000]. "The partial unification of domains, hybrids, and the growth of mathematical knowledge." *The Growth of Mathematical Knowledge*. Ed. by E. Grosholz and M. Breger. Dordrecht, Boston, and London: Kluwer Academic Publishers. 81–91.

Hamming, R.W. [1980]. "The unreasonable effectiveness of mathematics." *American Mathematical Monthly* **87**. 81–90.

Kleene, S.C. [1981]. "Origins of recursive function theory." *Annals of the History of Computing*. Vol. 3. 1. 52–67.

Lakatos, I. [1976]. *Proofs and Refutations: The Logic of Mathematical Discovery*. Ed. by J. Worrall and E. Zahar. Cambridge: Cambridge University Press.

Lautman, A. [1937]. "Essai sur l'unité des mathématiques et divers ecrits." Paris: Union générales d'éditions. Chap. Essai sur l'unité des sciences mathématiques dans leur developpement actual. 155–201.

— [1977]. *Essai sur l'unité des mathématiques et divers ecrits*. Paris: Union générales d'éditions.

Lax, P.D. [1986]. "Mathematics and its applications." *The Mathematical Intelligencer* **8** (4). 14–7.

Mancosu, P. [1996]. *Philosophy of Mathematics and Mathematical Practice in the Seventeenth Century.* New York and Oxford: Oxford University Press.

Mathias, A.R.D. [1992a]. "The ignorance of Bourbaki." *The Mathematical Intelligencer* **14** (3). 4–13.

— [1992b]. "What is Mac Lane missing?" *Set Theory of the Continuum.* Ed. by H. Judah. New York: Springer Verlag. 113–8.

Moore, E.H. [1910]. "Introduction to a form of general analysis." *The New Haven Mathematical Colloquium, Lectures Delivered before Members of the American Mathematical Society in Connection with the Summer Meeting Held September 5-8, 1906, under the Auspices of Yale University.* Vol. 2. American Mathematical Society colloquium publications. New Haven: Yale University Press.

Morrison, M. [2000]. *Unifying Scientific Theories: Physical Concepts and Mathematical Structures.* Cambridge: Cambridge University Press.

Putnam, H. [1967]. "Mathematics without foundations." *The Journal of Philosophy* **LXIV** (1). Reprinted in Putnam [1979]. 43–59.

— [1979]. *Mathematics, Matter and Method: Philosophical Papers.* 2nd ed. Vol. 1. Cambridge: Cambridge University Press.

Rav, Y. [1999]. "Why do we prove theorems?" *Philosophia Mathematica* **7**. 5–41.

Shapiro, S. [1991]. *Foundations without Foundationalism: A Case for Second-order Logic.* Oxford: Clarendon Press.

Stump, D.J. [2007]. "The independence of the parallel postulate and development of rigorous consistency proofs." *History and Philosophy of Logic* **28**. 19–30.

Wang, H. [1996]. *A Logical Journey: From Gödel to Philosophy.* Cambridge, MA: MIT Press.

Webb, J.C. [1980]. *Mechanism, Mentalism, and Metamathematics.* Dordrecht: D. Reidel.

Weyl, H. [1931]. *The Theory of Groups and Quantum Mechanics.* Trans. by H.P. Robertson. London: Methuen.

Wigner, E.P. [1960]. "The unreasonable effectiveness of mathematics in the natural sciences." *Communications on Pure and Applied Mathematics* **XIII**. 1–14.

ESSAY 2

Construction, Articulation, and Explanation
Phases in the Growth of Mathematics

Madeline Muntersbjorn

> Sie sind in der That in den
> Definitionen enthalten,
> aber wie die Pflanze in Samen,
> nicht wie der Balken im Hause
> (Frege [1884]: §88)

Venerable yet inconclusive disputes in the philosophy of mathematics suggest that what may be said of the whole of mathematics is relative to perspective and plural, rather than singular. Even so, some philosophers suggest that cognitive and contextual constraints have no necessary bearing on mathematical content. They distinguish the necessary facts of mathematics itself from the contingent features of mathematical practices that vary over time and place. For example, George and Velleman open *Philosophies of Mathematics* by suggesting that,

> The etiology of mathematical ideas, however interesting, is not something whose study promises to reveal much about the structure of thought: for the most part, the origin and development of mathematical ideas are simply far too determined by extraneous influences.
> (George and Vellemand [2002]: 2)

From the so-called maverick perspective, their lament in the conclusion is no surprise:

> Just as a seductive conception is presented which holds out the prospect of settling some of our original questions about mathematics, we discover . . . its leading idea is ultimately unsatisfactory or unrealizable. We are repeatedly drawn in, only to meet with disappointment. (Ibid.: 214)

The approach to the philosophy of mathematics advocated in this paper insists that the etiology of mathematical ideas is *totally relevant* to the nature of mathematical content. A careful study of the growth of mathematics—its origins and dispersion patterns—promises to reveal insight into the nature of mathematical kinds. A careful study of the pedagogy of mathematics—its explanations and heuristics—promises to reveal insight into the structure of thought. A less disappointing image of mathematics may require that we integrate divergent metaphysical insights together in a temporal dialectic. The thesis of the present paper is that different kinds of mathematical activities take place at different times corresponding to the recognition, representation and promulgation of mathematical content. This thesis may prove unsatisfactory for several reasons, depending on one's perspective. For example, some philosophers fear that pluralist approaches to the philosophy of mathematics substitute "an unclear and inconsistent amalgam for points of view which at least have the merit of a certain clarity" (Cohen and Nagel [1934]: ii). However, the general idea, that the growth of mathematics is best modeled via a dynamic dialectic, has been repeatedly realized in the history of philosophy of mathematics. This paper is a modern variation of an old theme, updated in light of recent developments in biology that inform what it means to say that *something grows*. In order to understand the growth of mathematical practices over time, we must learn to recognize how mathematical kinds exhibit life-cycle patterns in time.

From the perspective of the historiography of mathematics, the seminal idea of this paper may be far too simplistic. The approach to mathematics advocated here presumes that there is such a thing as mathematics, however elusive, and relates different kinds of mathematical practices and metaphysical perspectives inpost-mortem report on a particular kind of beast, a point we return to at the end of the paper a single abstract model. The very project of representing the growth of mathematics dismays some historians who specialize in particular loci in our past. Historians with an eye for detail question whether we should even try to understand mathematics as a whole, sensitive as they are to the subtle and swift shifts in what counts as "mathematics" over time. Like botanists dissecting particular specimens, they are so close to their subject matter that the etiology of mathematical kinds in general becomes insignificant. The advice given to botany students may resound with historians with a penchant for particulars: "[W]hen one begins to examine, in detail, all the morphological changes that occur from one phase to another ... the cycle itself may be ignored."[1]

In 1962 Heyting characterized the philosophy of mathematics as having achieved a cooperative, rather than competitive, tenor:

> The spirit of peaceful co-operation had gained the victory over that ruthless contest. No direction of research has any longer the pretension to represent the only true mathematics. The philosophical importance of research in the foundations of mathematics consists at least partly in the isolation of formal, intuitive, logical, and platonistic elements in the structure of classical mathematics, and the exact determina-

---

[1] The advice comes from online lecture notes written by Kent Simmons at the University of Winnipeg. URL: http://io.uwinnipeg.ca/~simmons/reproduc.htm. Last accessed: 10 February 2008.

tion of their scope and their delimitations.
(As cited by Franchella [1994]: 149)

Heyting reminds us to be wary of simplistic reductions of complex wholes into an arbitrary list of properties, especially those that privilege some particular feature over all the others. The present project is kin to proposals for pluralism in the philosophy of mathematics, even as it proposes a single abstract model for detailed consideration (cf. e.g., Davies [2005]). The model offered is but a preliminary sketch of how three kinds of mathematical practices relate to one another over time. Construction, articulation and explanation are phases in the growth of mathematics that correspond to the intuitive, formal, and platonistic on Heyting's scheme, respectively. There is no phase corresponding to logic because, on my view, each phase is *logical* in its own way. The diverse progeny born of the 20th century marriage of logic and mathematics teach us that there are many ways to systematize rational inquiry. For this reason, the number of distinct phases one could perceive in the growth of mathematics is unlimited, insofar as the number of periods one detects is a function of one's distance from the subject matter. The dialectic proposed is more like a generalized life-cycle model in botany than a post-mortem report on a particular beast or kind of species, a point we return to in conclusion.

Construction, articulation, and explanation are phases in the conception, cultivation, and dissemination of mathematical knowledge. The model respects the historians' insight that what "counts as mathematics" varies from one era to the next. Mathematical kinds, like biological species, are vigorous and robust in the now, somewhat stable over time, yet ultimately mutable, especially with respect to their relationships to one another. One can be a realist about mathematical kinds while insisting that their generation and proliferation does not proceed towards a pre-established, eternal, or inexorable goal. The model respects the philosophers' project of identifying the ties that bind diverse projects together as a particular kind of human activity, viz., mathematics. While the field of mathematics has grown in scope and depth, there are continuities in the kinds of activities people undertake that unite them as mathematical (even as they may be instances of other kinds of activities as well: measuring, sorting, modeling, etc.).

Construction, or the priority of perception and intuition, is the first phase. From Kant to Brouwer, the capacity to see with the mind's eye has been recognized as a mark of mathematical competence. On my model, mathematical intuitions are transformed into mathematical methods through the practices of construction, much as intuitionism suggests. Articulation, or the necessity of reason and representation, is the second phase. From Leibniz to Hilbert, the innovation of explicit and effective symbol systems that "do our thinking for us" has been recognized as a mark of mathematical progress. On my model, mathematical methods are expressed by mathematical demonstrations, broadly construed to include all phases of articulation from notational innovations to peer review, much as formalism suggests. Explanation, or the reality of content and understanding, is the third phase. From Plato to Pólya, we are reminded that teaching mathematics depends on a robust commitment to mathematical reality, i.e., wrong answers are not wrong simply because teacher says so but because *that's the way it is*. On my model, mathematical demonstrations, including heuristics, discovery stories, priority claims, and rival proofs, confer understanding through a process of explanation much as realism suggests.[2]

Mathematics inspires diverse metaphysical perspectives and replies precisely because its objects, concepts, and relations come into being gradually over time. Mathematical entities begin as private insights of persons or small groups that become public expressions of problem-solving strategies. When taught to the next generation, the referents of the representations employed in these techniques acquire degrees of objective autonomy in proportion to their overall intelligibility and usefulness. In other words, preference for one explanation over another is itself most readily explained in realist terms. A more viable proof does a better job of exhibiting clearly and precisely those features of

---

[2]Mathematical proofs have an unrivaled authority as if they come out of nowhere *ex nihilo* governing not only what is but what might be; yet even the most ironclad inferences turn out to have unexpected exceptions. Realists like Frege may be right about the universe insofar as mathematical laws play a decisive role in what can and cannot happen from one minute to the next. However, human ingenuity plays an important role in generating environments where these laws take concrete, and thus consequential, forms. For this reason, mathematics is neither created nor discovered, but rather cultivated through our interactions with one another and the world.

the entities responsible for the truth of the theorem such that this truth can be discerned independent of and distinct from the proof itself. In the next section, I discuss antecedent views from which this particular model descends. In the final section of the paper, I discuss each phase of the model in more detail.

1 Historical Considerations

The three phase model presented in this essay is a direct descendent of George Pólya's principle of consecutive phases:

> A first exploratory phase is closer to action and perception and moves on a more intuitive, more heuristic level. A second formalizing phase ascends to a more conceptual level, introducing terminology, definitions, proofs. The phase of assimilation comes last: ... an attempt to perceive the 'inner ground' of things, the material learned should be mentally digested, absorbed into the system of knowledge, into the whole mental outlook of the learner; this phase paves the way to applications on one hand, to higher generalizations on the other. (Pólya [1981]: 104)

For Pólya, this principle applies both to how individuals learn mathematics and how mathematics is cultivated by communities, that is, there is a close relationship between ontogeny and phylogeny in the growth of mathematics such that educators should adopt pedagogical strategies structured around this principle:

> For efficient learning, an exploratory phase should precede the phase of verbalization and concept formation and, eventually, the material learned should be merged in, and contribute to, the integral mental attitude of the learner.
> (Ibid.)

Pólya suggests that, in its most general form, his three-phase model applies to the development of science in general, not just mathematics: "In fact, in the historical development of various branches of science ... we may distinguish three phases" (ibid.: 133). However, a successful model of mathematics ought to capture important differences between mathematics and other disciplines as well as similarities. The analogy with life-cycle models is helpful at this point. All living organisms can be represented in a life-cycle model and both animals and plants reproduce sexually but only plants exhibit haploid and diploid phases characteristic of the alternation of generations. Perhaps the growth of all knowledge could be represented in some kind of life-cycle model, but we should expect mathematical knowledge to exhibit characteristic phases that distinguish it from other kinds of knowledge.

The model is also inspired by Saunders MacLane's questions and answers that appear at the end of his [1986] survey, *Mathematics Form and Function*. His discussion helps clarify the intended scope of the model offered in this paper. His first question is, what are the characteristics of a mathematical idea? He notes that mathematics starts from "an intuitive notion which gives guidance and purpose"; these "ideas arise not just from human activities or scientific questions; they also arise out of the urge to understand prior pieces of Mathematics" (ibid.: 415). Mathematical intuition does not begin exclusively with the raw sense data of human experience. More typically, mathematical intuition involves the recognition of implicit patterns inherent in extant mathematical practices. For this reason, the building blocks with which mathematical intuition constructs mathematical ideas are experiences of mathematical objects, properties, and relations.

MacLane's next question is, how does a mathematical form arise from human activity or scientific questions? His answer is that we "recognize and name an idea only after it has given rise to one or more formal expressions" (ibid.: 444). While this initial answer applies to science in general his subsequent discussion applies more narrowly to mathematics in particular: "It is the formal nature of Mathematics which makes it interpersonal, objective and exact" (ibid.: 454). While MacLane asks more than three questions in his discussion, the last question we consider in this paper is, how do we account for the unreasonable effectiveness of mathematics in providing models for science and knowledge? His answer sheds light on the third explanatory phase:

> By the choice of the successful cases of concept formation and deductions, various branches of Mathematics are se-

> lected and developed . . . [sometimes] supported by additional formal explanations of the regularity of new phenomena. (Ibid.: 446)

It is one thing to advocate an evolutionary epistemology of mathematics and another to specify, in detail, the selection pressures that determine the relative fitness of competing mathematical kinds.[3] On the model advocated here, the more fit articulations are those that facilitate explanations, in mathematics, science, and practical problem-solving contexts. In his paper of the same name Lakatos ([1978]: 61) asks, "What does a mathematical proof prove?" His approach to this question is to divide mathematical activity into three phases:

(1) pre-formal proofs;
(2) formal proofs; and
(3) post-formal proofs.

Thus, another way to understand the present project is to expand the scope of Lakatos' model by replacing "proofs" with "practices." With respect to construction or pre-formal practices, Lakatos notes, "There is indeed no respectable formal theory which does not have in some way or another a respectable informal ancestor" (ibid.: 62). Famously, Lakatos concludes that the articulation phase of mathematics is not exclusively concerned with the construction of proofs as concrete artifacts, but as conceptual activities or thought experiments. The formalist impetus to make the implicit features of our constructions explicit makes the transition from construction to articulation possible. The outcome of formalization is not abstract certainty, but improved understanding as part of the transition from articulation to explanation. Lakatos remarks:

> [The] problem is not to prove a proposition from lemmas or axioms but to discover a 'test-thought experiment' which creates the tools for a 'proof-thought experiment', which . . . instead of *proving* the conjecture *improves* it. (Ibid.: 96)

---

[3]The motivating insight of seminar paper by H. De Cruz (2007), "Universal Selection Theory and the Evolution of Mathematics." URL: http://ecco.vub.ac.be/~clement/seminars2007/ecco-decruz-14032007.pdf.

According to Lakatos ([1976]: 2), exclusive concern with the second phase, cut off from the other two phases, has led some philosophers to identify mathematical activity with proof articulation, e.g., Russell's jest that Boole's *Laws of Thought* was the first book ever written in mathematics. The distinction between proving and improving becomes more intelligible when we understand articulation as part of the whole rather than the entirety of mathematics. Articulation consists in those formal practices that bridge the gap between the intuitions of a select few and the problem-solving prowess of an entire community. With respect to the third phase, Lakatos ([1978]: 40) notes that: "The battle between rival mathematical theories is most frequently decided also by their relative explanatory power." This observation speaks to the question frequently posed in papers on explanation in mathematics, viz., if the goal of mathematics is certainty, why are so many results proven and re-proven again and again? The answer is that certainty is not the only goal: Competing proofs of the same or similar theorems seek to increase our understanding of the objects, properties, and, relationships invoked in an attempt to discern which are the most significant or, which properties of which objects "do the heavy lifting" in demonstrating why a particular result must be true.[4]

Both Frege and Lakatos agree that mathematics exhibits an organic unity that cannot be comprehended by a list of attributes. Frege says, when we study the "genuinely fruitful definitions of mathematics,"

> What we find there is not a series of associated characteristics; but rather a deeply, I would like to say, organic, interconnection of determinations. (Frege [2007]: 86)

Fruitful definitions in mathematics turn out to contain even more mathematical content than we supposed when they were first articulated. For this reason, they contribute to the growth of our knowledge. When mathematicians study what follows from their precepts, they do not simply take out of the box what they've put into it. Consequences are

---

[4]Cf. Tappenden [2008]: 14. The three-phase model is my response to his call for "a philosophical framework to help us keep the books while we sort it out," wherein "it" is the intricate relationship between definition and explanation (ibid.: 16).

contained in starting principles, not in the way that beams are contained in a house, but as plants are contained in seeds.[5]

Lakatos agrees that mathematics is distinct from our limited and contingent understandings of mathematics, but insists that mathematics grows as our understanding of mathematics changes in a dialectic that governs their interdependent realization. For Lakatos, mathematics "becomes a living, growing organism, that *acquires a certain autonomy from the activity which has produced it*":

> The genuine creative mathematician is just a personification, an incarnation of these laws which can only realise themselves in human action. Their incarnation, however, is rarely perfect. The activity of human mathematicians, as it appears in history, is only a fumbling realization of the wonderful dialectic of mathematical ideas. But any mathematician, if he has talent, spark, genius, communicates with, feels the sweep of, and obeys this dialectic of ideas.[6]
>
> (Lakatos [1976]: 146)

Mathematical activity produces new mathematical knowledge in such a way that our conclusions are organically related to our precepts as plants are to seeds. For Lakatos, the plant-seed metaphor is more apt and broader in scope than Frege intended. In particular, the complexity of the relationship between seeds and plants helps us to recognize complexity in the growth of mathematics. For one thing, plants come from seeds, which come from plants, which come from seeds, etc. Life-cycle models, with their arrows leading from one phase to the next, and then back to the beginning again, represent the dialectic Lakatos has in mind when he discusses the interplay between increased rigor and increased content in the growth of mathematics.[7] Further, the environment and other "extraneous" factors play decisive roles in the size, shape, and

---

[5] For more detailed discussion of §88, see Tappenden [1995].

[6] Lakatos' editors suggest that this passage reflects a young and overly Hegelian Lakatos. They rightly stress the indispensability and freedom of human ingenuity in determining a contingent mathematical future over the inevitable ideal imagined by Frege, inter alia.

[7] For more see Muntersbjorn [2007] in the volume of papers published from PMP2002.

arrangement of parts in a plant such that more than one possible plant is contained in a single seed. Similarly, mathematics could proliferate into more kinds than it probably will. *There's always more than enough past for any number of possible futures.*[8]

While Lakatos offers an open-ended view of mathematics, his disputant Gamma asserts that, "the growth of knowledge cannot be modeled in any language" to which Pi agrees. Does Lakatos agree? No, insofar as he (i) models the growth of mathematics in a complex dialog with several characters and elaborate footnotes and (ii) sees mathematics as a kind of knowledge. Yes, insofar as he warns against turning any one linguistic framework into a "conceptual prison" in the footnote to their exchange. Lakatos does not warn us away from model-building because he thinks there is such a thing as a "dialectic of discovery" that philosophers of mathematics should endeavor to explain. Rather, he warns against taking any one model as applicable to all kinds of knowledge and preferable to all others.

Dividing the growth of mathematics into phases does not commit us to an inevitable destiny or to the necessity of any particular phases. That is, the methodological approach to the philosophy of mathematics advocated here is historical, but not historicist—at least not in the sense attributed to "historicism" by Popper in his critique, *The Poverty of Historicism*:

> [T]he historicist method implies . . . that society will necessarily change but along a predetermined path that cannot change, through stages predetermined by inexorable necessity. (Popper [2002]: 50)

To which Lakatos replies,

> [T]he poverty of historicism is better than the complete absence of it—always providing of course that it is handled with the care necessary in dealing with any explosives.
> (Lakatos [1978]: 61)

Part of what it means to handle mathematical models of complex human practices with care is to remember that phases of development are individuated by the content accessible to us in our present understanding.

---

[8]Cf. Celluci [2000] and the "open world view."

For this reason, phases are not periods because they refer retrospectively and overlap significantly. Adopting a particular model of our past does not commit us to a particular future path. As Pickering reminds us,

> [M]odeling is an open-ended process with no determinate destination: a given model does not prescribe the form of its own extension . . . [T]he choice of any particular model opens up an indefinite space of different goals.
> (Pickering [1993]: 579)

Actual mathematical practices may be constitutive of a mature strain of X and nascent strains of $Y'-Y''''$ at the same time. As a result, phases in the history of mathematics have begun that we cannot discern because the relevant content has not yet been cultivated to the point of widespread accessibility. Making sense of this awareness requires a more detailed examination of the individual phases in the model, to which we now turn in the final section of this essay. In what follows, each phase is compared and contrasted with the divergent metaphysical views with which they are most closely associated.

2 A Particular Proposal

Construction Seeing with the mind's eye not only "what is" but, more importantly, "what can be done" to solve problems.

Articulation Making the implicit explicit by developing representations of problem-solving strategies that are correct, not just convenient or contingent.

Explanation Accounting for the success of these strategies by referring to the abstract objects and dependence relations responsible.

The first phase calls our attention to the cognitive origins of mathematical reasoning in an active perceptual process of seeing with the mind's eye not only "what is" but, importantly, "what can be done" to sort problems and seek solutions. Kant may be read (cf. Pólya) as offering a three-phase model of the growth of knowledge: First we develop intuitions, then conceptions, and finally ideas. Even though construction is an intuitive process, construction itself is not identical with the

20th century philosophical view known as intuitionism. For Brouwer the First Act of intuitionism is,

> Completely separating mathematics from mathematical language . . . recognizing that intuistionistic mathematics is an essentially languageless activity of the mind having its origin in the perception of a move of time. (Cf. Brouwer [1981]: 4)

It is one thing to acknowledge that mathematical reasoning engages non-verbal cognitive modalities and another to restrict the scope of mathematical content to the non-verbal perceptual for mystical reasons. The kinds of experiences that give rise to powerful mathematical intuitions are precisely the ones that can be articulated in mathematical languages later on and, for the working mathematician, most of her more mathematical experiences involve active engagement with extant mathematical languages and practices.

Construction entails ostensive signification and visual inspection insofar as it is a way of proceeding from what can only be seen with the minds' eyes to what can be placed before the eyes of another. Mathematical representations are not mere names of ideas even in algebra. As Kant notes, in the algebraic solution of problems, one puts

> symbols together in accordance with the form of notation for division, and thereby achieves by a symbolic construction equally well what geometry does by an ostensive or geometrical construction[.] (Kant [1998]: A717–B745)

Collins ([1974]) makes two points critically relevant to the phase of construction. First, algebra is an activity whereby individuals manipulate symbols in accordance with implicit methods as well as explicit rules. Second, one way to differentiate members of different intellectual communities is by reference to their distinct, yet unexpressed, differences with respect to what may or may not be performed in a given context:

> Learning algebra consists of more than the memorization of sets of formal rules; it involves also knowing how to *do* things . . . Now, an important difference between members of different paradigm groups . . . lies in the contents of their tacit understandings of the things that they may legitimately

do with a symbol or a word or a piece of apparatus.

(Ibid.: 167–8)

The analogy between the phases of growth in mathematics and life-cycle models is helpful again for as Kent Simmons advises his botany laboratory students:

> Too often life cycles are examined as a series of somewhat unrelated static events. Ideally, they should be considered as a series of phases in the development of the organism. Each phase is simply a transition from the previous to the subsequent phase.[9]

Similarly, construction is not limited to rudimentary perceptual or non-verbal activities, but includes only those sophisticated perceptual and non-verbal practices that make the transition from old mathematical languages to new ones possible. Kant recognizes the dynamic interplay between intuition, signs, and truth. He notes how intuition operates on signs and how formal expressions ultimately contribute to the rigor of our reasoning as well as the truth, which comes with but is distinct from the proof:

> Even the way algebraists proceed with their equations, from which by means of reduction they bring forth the truth together with the proof, is not a geometrical construction, but it is still a characteristic construction, in which one displays by signs in intuition the concepts, especially of relations of quantities, and, without even regarding the heuristic, secures all inferences against mistakes by placing each of them before one's eyes.[10]   (Kant [1998]: A734–B762)

We should also contrast construction with idealism, the view that nothing exists besides the thoughts of thinking beings, as well as its cousin, nominalism, the view that nothing mathematical exists besides

---

[9] URL: http://io.uwinnipeg.ca/~simmons/reproduc.htm. Of the generalized plant life-cycle model, Simmons notes, "surprisingly, it actually applies to almost all plants."

[10] For assistance unpacking this dense passage see Shabel [2003].

the names of non-existent entities created by thinking beings. Construction practices are not creative processes we undertake because we do not have access to mathematical reality; rather, these practices are the means whereby intellectual access to mathematical reality is obtained. By interacting with mathematical symbols we become acquainted with their referents such that, over time, we we learn about real things that are distinct from their appearances. Or at least this is how I interpret Kant's rejection of idealism in the *Prolegomena*:

> Idealism consists in the claim that there are none other than thinking beings; the other things that we believe we perceive in intuition are only representations in thinking beings, to which in fact no object existing outside these beings correspond. I say in opposition: There are things given to us as objects of our senses existing outside us, yet . . . we are acquainted only with their appearances, i.e., with the representations that they produce in us because they affect our senses. (Kant [1997]: §4:289)

On this interpretation, we experience mathematical reality when we employ a suitable system of representation for making our mathematical reasoning explicit. Mathematical statements represent our thought patterns (themselves representations of prior sense experiences) and may yet be present to us as objects of our sense.

During the second phase, articulation, the security of our inferences is subject to the closest scrutiny. Mathematical reasoning makes the implicit explicit by cultivating concrete means to represent problem-sorting and solution-generating strategies as necessarily correct, not just convenient or contingent. Articulation is not to be confused with the rationalism one finds in Leibniz in translation: "All our thinking is nothing but the connection or substitution of signs, whether these are words or marks or images" (Westerhoff [1999]: 453). Articulation is, however, closely related to Leibniz's understanding of *Ratiocinatio* as a particular kind of reasoning distinct from the perceptual or non-verbal. In an unpublished manuscript, Leibniz ([1965]: 31) writes: "Omnia Ratiocinatio nostra nihil aliud est quam characterum connexio et substitutio, sive ille characteres sint verba, sive notae, sive denique imagines."

Although articulation is concerned with the formal expression of mathematics, articulation should not be confused with the 20th century view known as formalism. Hilbert envisioned a proof theory that would

> dispose of the foundational question in mathematics once and for all by turning every mathematical statement into a concretely exhibitable and rigorously derivable formula and thereby transferring the whole complex of questions into the domain of pure mathematics. (Hilbert [1998]: 228)

Articulation not only translates our insights but also transforms our intuitive faculties. It is perhaps surprising to discover just how much human cognitive endowments shape the topology of mathematics, since we have been conditioned to see it as somehow otherworldly in origin. But few realize just how much the topology of mathematics shapes the brains we develop when we grow up in measurable worlds. Polanyi ([1962]: 70) notes that the growth of our intuitive faculties is nurtured by an increase in our expressive capacities: "[I]ndeed our mute abilities keep growing in the very exercise of our articulate powers." The more we learn how to say what we know, the more we learn to recognize new things that we cannot articulate, at least not yet.

As pointed out earlier in the discussion of Lakatos, articulation is a creative process that improves as it proves and, as it turns out, both mathematics and the mind undergo transformation. For this reason, Tiles notes,

> [T]he line between invention and discovery cannot easily be made sharp ... because discovery includes discovery of ways of doing things as well as of the existence of previously unknown things, and discovering new ways of doing things frequently involves inventing new instruments.
> (Tiles [1989]: 3)

Her point helps us to see why the model offered here is a 3-phase model. The first two phases, construction and articulation, are close cousins of the context of discovery and the context of justification, respectively, two earlier phases in an enduring model from earlier times. The present model has more than two phases so that it undermines the mind/math dualism inherent in the discovery/justification model. The model has three phases rather than some larger number, for the sake of simplicity; recall the point earlier regarding the unlimited number of possible phases one could ascribe to the growth of mathematics. Explanation, or the process whereby successful articulations are taken to refer to concepts, objects, and relations, is not the final or ultimate stage, but the transitional phase between articulation and construction.

Explanation starts with the awareness that some formal articulations are better than others (even if none of them are, in fact, universal or necessary in all possible worlds). The success of the best explanations in mathematics is itself explained by their power to make shared mathematical understanding possible. Explanations account for successful articulations by invoking the abstract objects and dependence relations most directly responsible for their truth. Good explanations encourage us to embrace realism, but do not commit us to become platonists as understood during the 20th century:

> Mathematical platonists usually insist that mathematical entities are abstract in not being spatio-temporally located and thus in not having any causal powers.
> (George and Vellemand [2002]: 7)

However, explanation as intended here is closely related to Plato's metaphysics, understood not as an ontological account of what is, but as a pedagogical narrative about how mathematics is learned and why philosophers ought to study how people learn mathematics. In the *Meno*, Socrates leads a young uneducated slave to the right answer to a mathematics problem, not by telling him how to solve it, but by correcting his mistakes until he gets the right answer. After his initial success, Socrates observes:

> These opinions have now just been stirred up like a dream, but if he were repeatedly asked these same questions in various ways, you know that in the end his knowledge about these things would be as accurate as anyone's.
> (Plato [1976]: §85d)

Socrates calls our attention to mathematical practices as something one can do repeatedly, as an unconscious habit or as part of an acquired tradition. Mathematics involves complicated languages wherein the same questions can be asked in various ways—some more illuminating and suggestive than others. Mathematical knowledge is acquired in time—it takes a while for the practices to take hold. In the end, the mathematical learning activities one undertakes matter more than one's birthright, for anyone who achieves any mathematical accuracy at all achieves the very same amount of accuracy as anyone else.[11]

McLarty ([2005]) interprets Plato's divided line in *The Republic* as a four-faculty model wherein imagination, belief, thought, and understanding represent different cognitive processes involved in coming to know. Mathematical knowledge is not the goal of dialectic—mathematical knowing is an intermediary that illustrates how we get from mere belief to genuine understanding (ibid.: 121ff). McLarty shows how Socrates teaches Glaucon not to value math as a superior kind of knowledge but as an illustrative example of how knowledge comes about:

> Socrates says the philosopher's goal is not to learn mathematics but to learn what mathematics is, and 'if all our investigation of these studies brings out their community with each other and common origin and implies that they are kin then our hard work has advanced our goal[.]'   (Ibid.: 127)

To study 'what mathematics is' is not to practice mathematics. Rather, the goal of the philosopher is to articulate the organic connections between various heterogeneous practices and specify how they all relate to one another as briefly yet precisely as possible. The "hard work" that this project requires is ironic understatement. But the good news is that progress can be made. When we are able to establish genealogical relationships between kinds of mathematical practices, we do not labor in vain.

Explanation is distinct from construction, for it is a pedagogical process that generates understanding by appeal to first principles rather than direct experience. Grabiner notes the importance of this distinction:

---

[11] Of course, there are good reasons to suppose that, contrary to this universality of accuracy, degrees of competence, and kinds of evidence turn out to be very relevant to the dispersion of mathematical ideas.

> Teaching . . . stimulated mathematicians to consider the foundations of their subject. In presenting a subject like analysis to beginners, no appeal could be made to the way the concept is understood in use, since the beginner did not have the experience needed . . . Having students tends to force a teacher to expound the first principles of a subject clearly and to think those principles through anew.
> (Grabiner [1981]: 25)

An understanding of how explanation works in mathematics helps us cultivate a more precise distinction between objective and object. *Articulations* are objective in the sense of interpersonal and accessible: "Representing the thought-things by symbols or signs is precisely what makes the thought things objective" (Carter [2004]: 255 on Hilbert). *Explanations* invoke mathematical objects as independent of any system of signification or thought. Symbols for operations may be objective before they refer to any actual objects. Deciding which symbols refer to actual objects is a large, and largely contested, part of the explanatory phase of mathematical practices, especially when one construes "mathematician" to include teachers, engineers, and scientists whose work depends on the robust stability of mathematical kinds (e.g., counting numbers).

Realism about certain kinds of mathematical content may not be metaphysically necessary but may yet be necessary for pragmatic reasons. On my view, practically necessary is necessary enough. Select symbols refer to mathematical objects in the explanatory phase in such a way that learning new mathematics becomes a process of discovery. Understanding objects as such enables mathematicians to see what they might also do with these things in addition to what may be merely affirmed of them and, for the time being, completes the cycle.

When we adopt an emergent realist approach to mathematical content we are able to transition from explanation back to construction via enhanced capacities of perception. In their paper presented at the first PMP conference, Van Kerkhove and Myin ([2002]: 361) contend, "for one committed to take mathematical practice seriously, mathematical objects and properties should find a place *out there* in the *real* world." Inspired by Gibson's ecological view of perception, they do not ask how mathematicians have access to objects beyond their ken. Instead,

> We would say that mathematical entities or properties can be *directly perceived* by those perceivers familiar with the constitutive operations connected with the entities or properties. (Ibid.: 365)

During the transition from explanation to construction, the growth of mathematics may be characterized as "the growth of ontology through the extension of practice" (ibid.: 368). When asked why a contemporary philosopher of mathematics should take objects seriously, especially since "pure mathematics" can be done without explicitly invoking any objects at all, my answer is that a philosopher of mathematics must be able to distinguish between objects and operations in order to recognize the dialectic whereby mathematics grows over time.

An ancient parable presents blind men who study an elephant from different points who come to different conclusions about what "the elephant" is like. The man at the side thinks the elephant is like a wall, the man at the tail says the elephant is like a rope, while the man who touches the ear urges the elephant is like a fan, etc. A particularly old version from the Han dynasty concludes:

> How they argued! Each one insisted that he alone was correct. Of course, there was no conclusion for not one had thoroughly examined the whole elephant. How can anyone describe the whole until he has learned the total of the parts?[12]

The study of mathematical practices, broadly construed, considers all of the parts of mathematics as relevant to the philosophy of mathematics—logical as well as historical, social, and cognitive dimensions. Exclusive concern for foundations may be likened to groping a tusk and marveling at how like unto a spear the elephant is. Cut off from the rest of the elephant, tusks lose their original meaning and become a decorative material we bend to our will: pretty but pricey. While the entire elephant

---

[12] URL: http://www.noogenesis.com/pineapple/blind_men_elephant.html. Versions of this parable hail from India, China, and Africa. Famously rendered into English by American poet John Godfrey Saxe (1816–87) in the mid-19th century. An excellent rendition in granite is *Blind Men Study the Elephant* by Liu Zhangde (Wuhan; 1985). For photo see John T. Young's web page, *Public Art in China*. URL: http://faculty.washington.edu/jtyoung/chinaart.html.

may elude our best efforts, we could try to understand them as more than our best source of *ivory*.[13]

The logical structure of mathematics is unrivaled in perspicacity and persuasive power. But when foundational inquiry is isolated from the rest of mathematical practice such that the psychology and pedagogy of mathematics are dismissed as irrelevant to *mathematics itself,* one enters a dubious realm where what matters the most are aesthetic choices and technical skills. Proofs isolated from the rest of mathematical practices lose their meaning and power. While the complex heterogeneity of mathematics in its entirety will elude our best efforts, we can understand mathematics as more than a source of inferential consequences or necessary truths.[14]

However apt we find the comparison between mathematics and the elephant in the parable, I conclude by stressing two vitally important differences. For the men in the story do not have direct access to their subject and the elephant is imagined as a delimited whole capable of being divided into parts: ear, leg, tail, etc. In contrast, mathematics is a subject we can experience directly, albeit with training and effort. Further, mathematics is not a fixed whole, but a diverse collection of practices, the very nature of which is to grow beyond their present boundaries. For these reasons, the relationship between mathematics and mathematicians is more like the relationship between plants and their seeds than that between the elephant and the blind men.

Bibliography

Brouwer, L.E.J. [1981]. *Brouwer's Cambridge lectures on intuitionism.* Ed. by D. Van Dalen. New York: Cambridge University Press.

---

[13]The parable was expressed on a much smaller scale by Robert Olszewski in a 1″ resin miniature by Goebels for collectors in 1986—reminding us in a faux netsuke nutshell that certain features of an elephant can be rendered in ivory just as particular features of mathematics can be modeled in formal proofs.

[14]The likeness between the elephant in the parable and mathematics inspires Dan Kalman to write *The Mathematical Elephant,* an e-column in American University's *Discrete Observer.* I especially like how he marks the belly of the beast with a big question mark: http://www.american.edu/academic.depts/cas/mathstat/do.cfm.

Carter, J. [2004]. "Ontology and Mathematical Practice." *Philosophia Mathematica* **12**. 244–67.

Celluci, C. [2000]. "The Growth of Mathematical Knowledge: An Open World View." *The Growth of Mathematical Knowledge*. Ed. by E. Grosholz and H. Breger. Dordrecht: Kluwer. 153–76.

Cohen, M.R. and E. Nagel [1934]. *An Introduction to Logic and Scientific Method*. New York: Harcourt and Brace.

Collins, H.M. [1974]. "The TEA Set: Tacit Knowledge and Scientific Networks." *Social Studies of Science* **4**. 165–86.

Davies, E.B. [2005]. "A Defence of Mathematical Pluralism." *Philosophia Mathematica* **13**. 252–76.

Franchella, M. [1994]. "Heyting's Contribution to the Change in the Research in the Foundations of Mathematics." *History and Philosophy of Logic* **15**. 148–72.

Frege, G. [1884]. *Die Grundlagen der Arithmetik*. Breslau: Wilhelm Koebner.

— [2007]. *The Foundations of Arithmetic*. Trans. by D. Jacquette. First published in 1884. New York: Pearson Longman.

George, A. and D.J. Vellemand [2002]. *Philosophies of Mathematics*. Oxford: Blackwell.

Grabiner, J. [1981]. *The Origin's of Cauchy's Rigorous Calculus*. New York: Dover.

Hilbert, D. [1998]. "Problems of the Grounding of Mathematics." *From Brouwer to Hilbert: The Debate in the Foundations of Mathematics in the 1920s*. Ed. by P. Mancosu. Trans. by Note = "First published in 1929 Mancosu P.". New York: Oxford University Press. 227–33.

Kant, I. [1997]. *Prolegomena to Any Future Metaphysics*. Trans. by G. Hatfield. First published in 1783. Cambridge University Press.

— [1998]. *Critique of Pure Reason*. Trans. by P. Gruyer and A.W. Wood. First published in 1781–87. Cambridge: Cambridge University Press.

Lakatos, I. [1976]. *Proofs and Refutations: The Logic of Mathematical Discovery*. Ed. by J. Worrall and E. Zahar. Cambridge: Cambridge University Press.

— [1978]. "What does a mathematical proof prove?" *Philosophical Papers.* Vol. 2. First published in 1961. Cambridge University Press. 61–9.
Leibniz, G.W. [1965]. *Die Philosophischen Schriften.* Ed. by C. Gerhardt. Vol. 7. Hildesheim: Georg Olms.
MacLane, S. [1986]. *Mathematics Form and Function.* New York: Springer Verlag.
McLarty, C. [2005]. "'Mathematical Platonism' *Versus* Gathering the Dead: What Socrates teaches Glaucon." *Philosophia Mathematica* **13**. 115–34.
Muntersbjorn, M. [2007]. "Mathematical Progress as Increased Scope." *Perspectives on Mathematical Practices.* Ed. by B. Van Kerkhove and J.P. Van Bendegem. Dordrecht: Springer. 107–17.
Pickering, A. [1993]. "The Mangle of Practice: Agency and Emergence in the Sociology of Science." *The American Journal of Sociology* **99**. 559–89.
Plato [1976]. *Meno.* Trans. by G. Grube. Indianapolis: Hackett Publishing Company.
Polanyi, M. [1962]. *Personal Knowledge: Towards a Post-Critical Philosophy.* Chicago: University of Chicago Press.
Pólya, G. [1981]. *Mathematical Discovery: On Understanding, Learning, and Teaching Problem Solving.* Vol. II. First published in 1965. New York: John Wiley & Sons.
Popper, K. [2002]. *The Poverty of Historicism.* First published in 1957. London: Routledge.
Shabel, L. [2003]. *Mathematics in Kant's Critical Philosophy: Reflections on Mathematical Practice.* New York: Routledge.
Tappenden, J. [1995]. "Extending Knowledge and 'Fruitful Concepts': Fregean Themes in the Foundations of Mathematics." *Noûs* **29**. 427–67.
— [2008]. "Mathematical Concepts and Definitions." *The Philosophy of Mathematical Practice.* Ed. by P. Mancosu. Oxford University Press.
Tiles, M. [1989]. *The Philosophy of Set Theory.* New York: Dover.
Van Kerkhove, B. and E. Myin [2002]. "Direct Perception in Mathematics: A Case for Epistemological Priority." *Logique et Analyse* **179-80**. 357–72.

Westerhoff, J. [1999]. "Poeta Calculans: Harsdorffer, Leibniz, and the mathesis universalis." *Journal of the History of Ideas* **60** (3). 449–67.

ESSAY 3

Laplace, Lakatos, and Luck
Accommodating Extrinsic Justification in Quasi-Empiricism

Ian Dove

Lakatos divided theories along two dimensions: type of truth-value injected into a theory, and the characteristic direction, or *flow*, the truth-value takes upon injection. He explicitly distinguished three theory types (or programmes) along these two[1] dimensions: Euclidean, Em-

---

[1] It may be tempting to think that Lakatos was more concerned with the direction of the flow, which is determined by the point at which *some* value gets injected, rather than the values themselves.

> We can get a long way merely by discussing *how* anything flows in a deductive system without discussing the problem of *what in fact flows* there, infallible truth or only, say, Russellian 'psychologically incorrigible' truth, Braithwaitian 'logically incorrigible' truth, Wittgensteinian

Direction ↘ / Value →	Truth	Falsity
Top Down	Euclidean	*Fictionalist?*
Bottom Up	Inductivist	Quasi-Empiricist

Table 1: Lakatosian Theory Types

piricist, and Inductivist—presented in Table 1.[2] The table contains a number of simplifying assumptions. For example, the table assumes bivalence. Also, the table assumes that theories shall be distinguished by a unique and single direction. This means that there are no theories in which a truth-value flows both up and down. Rather, within any single theory type, truth-value flows in one and only one direction. A Euclidean theory has *truth* injected at the top: the level of axioms or postulates. From this injection, truth flows down to the theorems or consequences. Since the conduit is deduction, and deduction is truth-preserving, the truth of the axioms guarantees the truth of the theorems. This is the orthodox view of justificatory practice in mathematics.[3] It is against this view that Lakatos directs his strongest criticism. The

'linguistically incorrigible' truth or Popperian corrigible falsity and 'verisimilitude', [or] Carnapian probability. (Lakatos [1978b]: 6)

The phrase *how anything flows* gives voice to this temptation. However, Lakatos clearly envisions a characteristic truth-value for each theory type as well as a characteristic direction for the flow of that value. Whenever he discusses the theory types he always demarcates the theories in terms of both value and direction. Even the values presented in the above quote are, with the possible exceptions of versimilitude and Carnapian probability, truth-values. Later in the paper I consider Lakatos's apparent focus on directionality.

[2] Although Lakatos mentions and quickly dismisses the idea of injecting falsity at the top (cf. Lakatos [1978b]: 4), I've included the title *Fictionalism* as a placeholder for this possibility. Whatever inferences compose this possibility, they cannot be, for example, *ex falso quodlibet*, $\{\phi \land \neg \phi\} \vdash \psi$, as this is neither truth-preserving nor falsity-preserving from the top down. Instead, it shows that any deductive system for which *ex falso quodlibet* is an accepted rule of inference and for which falsity is injected at the top may be vacuous.

[3] Hereafter, I shall refer to this as simply *mathematical practice*. This shouldn't cause confusion insofar as Lakatos was concerned to explain mathematical knowledge in terms of the justificatory practice of mathematics, i.e., by explicitly rejecting Euclidean theories of mathematics in favor of a quasi-empiicist account.

criticism is in the form of *rationally reconstructed* episodes from the history of mathematics that show Euclidean orthodoxy to be at odds with mathematical practice.

Against Euclidean orthodoxy, Lakatos argues that mathematical practice is quasi-empirical. This can be understood as the application of Popper's falsificationist methodology to mathematics. By definition, a quasi-empiricist theory has only *falsity* injected into the system at the bottom. Thus, if mathematics is quasi-empiricist, then mathematical propositions are either refuted or corroborated, though never established conclusively.

That Lakatos's quasi-empiricism is an antidote to Euclidean orthodoxy is oft-discussed. In this paper, though, I investigate the mostly neglected *inductivist* methodology and its possible instantiation in the history of mathematics. My guide will be Lakatos, initially, but I will draw some terminology from Penelope Maddy's ([1988a]; [1988b]; and [1992]) account of philosophy of mathematics. To forestall possible terminological confusion, let me make several definitions explicit. First, *inductivism* is a theory type and not a conduit through which a truth-value can flow. It is tempting to identify inductivism with inductive inferences. Though this would be an error. Within all of Lakatos's theory types, the conduit or channel through which a truth-value flows is deductive—even in inductivism. This also explains why there are two, rather than three, dimensions along which to demarcate the theory types. What Lakatos calls *inductivism* could be termed *abductivism*. The distinguishing features of inductivism are that Truth is injected at the bottom from which it flows upward. Thus, inductivism is distinguished from quasi-empiricism in virtue of the characteristic value injected into the system—Truth for inductivism and Falsity for quasi-empiricism. They share the same characteristic direction for the flow of the value: upwards. Furthermore, inductivism is distinguished from Euclideanism in terms of the direction of the flow of value within the system: upwards for inductivism and downwards for Euclideanism; though the same truth-value, Truth, flows within each of these theory types.

From at least 1760 through 1826, mathematicians manipulated divergent series (almost) with impunity. In 1760, Euler published a paper on divergent series in which he *summed* a divergent series in numerous different ways. Although other mathematicians had *summed* divergent series before this, Euler's paper makes a nice starting point from which to investigate several methods of manipulating such series. An obvious endpoint for this period of divergent mania is 1826 when both Abel and Cauchy rejected the use of divergent series. Abel's pithy dismissal resonates: "[D]ivergent series are the work/invention of the devil." (Abel [1964]: 257–8) This proclamation, along with a similar claim by Cauchy, slowed work employing divergent series. Yet, before the Abel/Cauchy banished divergent series from rigorous mathematics, Euler in 1760, and Laplace from 1798–1821 justified their use of divergent series not by *rigorous* proof, but by what Lakatos would call an inductivist methodology or what Maddy would call *extrinsic* justifications.[4] Likewise, astronomers continued to use divergent series unaffected by the claims of Abel and Cauchy. Even Cauchy continued working with and on divergent series (cf. Kline [1972]: 974). DeMorgan went even further in arguing for the use of divergent series.

> [T]he most determined rejector of all divergent series doubtless makes . . . use of them in his closet.
>
> (Quoted in Kline [1972]: 1097)

Perhaps surprisingly, almost 100 years after Laplace used divergent series to make accurate predictions in positional astronomy, Poincaré rigorously proved the correctness of series manipulations like Laplace's. I argue below that Laplace justified his use of divergent series in accord with inductivism. Yet, insofar as Lakatos rejects inductivism on the grounds that truth doesn't flow upwards in a system of deductive relations, Laplace was merely epistemologically lucky. Since deduction isn't truth-preserving from the bottom up, that is, from consequence to premise, having a true consequence is no guarantee that the premise or premises will be true. That Laplace's premises turn out to be true, then, is a matter of luck and not principle. In this paper I suggest that

---

[4]Later, I consider whether extrinsic justification is best understood in terms of heuristic. There are good reasons for thinking that it is, but these are not decisive. Rather, given Maddy's ([1988a] and [1988b]) litany of extrinsic justifications, there are reasons for thinking that extrinsic justification are precisely ruled out by Lakatos's criticism of Gödel (see below).

inductivism is both a normal part of justificatory practice, and that it can be incorporated into a nearly Lakatosian account.

1 Lakatos *on* Inductivism

Lakatos criticizes the orthodox, foundational view of mathematical practice. In its place he offers quasi-empiricism. What is not as widely discussed is Lakatos's dismissal of *inductivism*. Recall from Table 1 above that inductivism has *truth* injected at the bottom, at the level of consequences, and it *flows* upwards to the axioms or postulates. This view of *inductivism* does not fit perfectly into the standard deductive/inductive dichotomy. For Lakatos the distinction rests solely upon the direction that an injection of truth would take upon injection into a system.

> The Inductivist Programme was a desparate effort to build a channel through which truth flows upwards from the basic statements, thus establishing an additional logical principle, *the principle of retransmission of truth.*
>
> (Lakatos [1978b]: 7)

The difficulty for inductivism is that the only conduit Lakatos imagined for the transmission of truth-value is deduction. Deduction is truth-preserving from the top down, and falsity preserving from the bottom up. Lakatos found the truth-preservation—or retransmission—from the bottom to be impossible.

> An inductive channel did not look so obviously impossible in the seventeenth century as it does today[.]
>
> (Lakatos [1978b]: 7)

Many mathematicians and philosophers of mathematics are cited by Lakatos as endorsing an inductivist methodology in mathematics. Lakatos quotes the following passage as typical of Gödel's *inductivist* leanings.

> There might exist axioms so abundant in their verifiable consequences, shedding so much light upon the whole discipline, and furnishing such powerful methods for solving given problems that quite irrespective of their intrinsic necessity they

would have to be assumed at least in the same sense as any well established physical theory.

(Gödel [1947]: 521, quoted in Lakatos [1978a]: 26)

The important feature of Gödel's claim is that the axioms garner support from the verification of their consequences. That is, Gödel explicitly accepts the kind of *bottom up* justification that Lakatos finds 'obviously impossible.'

It is important to keep in mind that for Lakatos *inductivism* is as unsatisfactory an account of mathematical practice as Euclideanism.

> The gravest danger then in modern philosophy of mathematics is that those who recognize the fallibility and therefore the science-likeness of mathematics, turn for analogies to a wrong image of science. The twin delusions of 'progressive intuition' and of induction can be discovered anew in the works of contemporary philosophers of mathematics.
>
> (Lakatos [1978a]: 42)

The first delusion, viz. 'progressive intuition,' is emblematic of Euclideanism. Recall that the Euclideans want to found mathematical practice upon the search for intuitively obvious axioms. The delusion of induction, at least for Lakatos, stems from the lack of a logical conduit or channel through which truth can move from consequence to premise. This is surely one of the closest ties between Lakatosian and Popperian methodologies.

> Science is not a system of certain, or well-established, statements; nor is it a system which steadily advances towards a state of finality. Our science is not knowledge (epistēmē): It can never claim to have attained truth, or even a substitute for it, such as probability.
>
> Yet science has more than mere biological survival value. It is not only a useful instrument. Although it can attain neither truth nor probability, the striving for knowledge and the search for truth are still the strongest motives of scientific discovery. (Popper [2002]: 306)

In the case of either Popper or Lakatos, the standard conduit through which truth-value flows is deduction. Since deduction cannot carry truth upwards, any purely inductivist confirmation will be lucky rather than rational. Again, the truth of the consequences of an argument do not entail the truth of the premises of that argument.

2 Recent Inductivism: Maddy's Extrinsic *Justifications*

In Penelope Maddy's [1992] discussion of realism in mathematics, she claims that there are two kinds of justification in mathematics: *intrinsic* and *extrinsic*.[5] Intrinsic justification is simply self evidence or 'intuitiveness.' Hence, Maddy, insofar as she accepts intrinsic justification in mathematics is a Euclidean in Lakatos's sense.

On the other hand, Maddy also allows for *extrinsic* justifications in mathematics. Let us say that a thesis, conjecture or axiom is supported or justified *extrinsically* insofar as its consequences are verified by some method or source unrelated to the original means by which the consequences are drawn. From this support of the consequences, one draws support for the methods that initially resulted in these consequences. This is insufficient to cover the totality of extrinsic supports that Maddy allows (see Maddy [1988a] and [1988b]). Still it gives the flavor of extrinsic justification.

In support for her thesis that extrinsic justification is a standard method in mathematics, Maddy twice quotes the passage from Gödel that Lakatos cited as delusional (Maddy [1992]: 32–3, 77–8 and Lakatos [1978a]: 26), as well as an interesting passage in which Gödel identifies mathematical verification with fruitfulness or success.

> [B]esides mathematical intuition, there exists another (though only probable) criterion of the truth of mathematical axioms, namely their fruitfulness in mathematics and, one may add, possibly also in physics.
> (Gödel [1983]: 485, quoted in Maddy [1992]: 77)

---

[5]I leave to one side Maddy's ([1988a] and [1988b]) discussion of a third category: *rules of thumb*.

The curious element of this quote is the final conjunct. This is inductivism in a very radical sense in as much as the injection of truth into mathematics comes by way of mathematics' fruitfulness in physics. But it isn't non-sensical in the way implied by Lakatos when he claimed that the inductive route is delusional. Although Lakatos can't imagine a conduit through which truth could climb from consequences to the level of axioms, concepts, or conjectures, many philosophers are convinced that such a conduit exists in science. Pierce, for example, distinguishes deduction, induction, and *abduction* or what we now call *Inference to the Best Explanation* (IBE). In the next section I examine two apparent cases of abduction in the history of mathematics.

3 Divergent Series: The Devil's Work?

There was a time when divergent series were standard mathematical fare. Such luminaries as Bernoulli and Leibniz manipulated divergent series to great effect. Grandi, Bernoulli, Leibniz, and Euler, for example considered a famous series with an infamous sum.

$$s = 1 - 1 + 1 - 1 + 1 - \cdots = \frac{1}{2} \tag{1}$$

The resulting sum can be determined in several ways. Both Leibniz and Bernoulli, for example justified it as a probability argument. To understand this version, notice that one can give different groupings of the terms of the series. On one grouping the terms cancel completely, giving the sum 0. On another, the terms cancel nearly completely, leaving the sum 1.

$$s = \underbrace{1-1}_{0} + \underbrace{1-1}_{0} + \underbrace{1-1}_{0} + \underbrace{1-1}_{0} + \cdots = 0$$

$$s = 1 \underbrace{-1+1}_{0} \underbrace{-1+1}_{0} \underbrace{-1+1}_{0} \underbrace{-1+1}_{0} - \cdots = 1$$

By taking the *average* of these two *sums*, Bernoulli concluded that the original sum is given in (1). This method should be dismissed because even with convergent series it is not (always) possible to rearrange terms without altering the sum. If instead, the value is obtained by taking the average of the partial sums, then there is nothing objectionable about this method except for the oddity of the series under consideration.

Grandi (cf. Kline [1972]: 445–6) reached the same result by a different method (and this is similar to a method that Euler used). He let $x = 1$ in the following equation.

$$\frac{1}{1+x} = 1 - x + x^2 - x^3 + \cdots$$

Grandi's argument requires that an equivalence already be in place for the substitution to work. As such equivalences aren't available for every non-convergent series, different methods are required. And this is why Euler's systematic considerations of this problem allow him to proceed with generality where previous work proceeded with particular sums.

3.1 EULER

Euler's ([1760]) paper *De seriebus divergentibus*, though it was written, perhaps, 14 years earlier, contains an interesting *sum* of a divergent series.[6] Begin with the following series.[7]

$$w = x - (1!)x^2 + (2!)x^3 - (3!)x^4 + (4!)x^5 - \cdots$$

By setting $x = 1$ we get the following.

$$w = 1 - 1 + 2 - 6 + 24 - 120 + 720 - 5040 + \cdots$$

Because this is an alternating series, Euler first considers alternating series in general.

$$A = a - b + c - d + e - f + \cdots$$

By considering the series as a sequence and neglecting the signs, one can extract what Euler calls "the first difference of the series."

$$(b-a), (c-b), (d-c), (e-d), \ldots$$

---

[6]After drafting and presenting an earlier version of this paper I learned of a slightly different, but illuminating discussion of Euler's method, Sandifer [2006]. I've corrected several errors in my original presentation on the basis of Sandifer's explanations.

[7]Kline ([1972]: 451) points out that this series satisfies the differential equation $x^2 y' + y = x$. From this Euler is able to take the series as an expansion of the resulting (integral) equation.

From this one can extract the second difference.

$$\underbrace{(c-b)-(b-a)}_{c-2b+a}, \underbrace{(d-c)-(c-b)}_{d-2c+b}, \underbrace{(e-d)-(d-c)}_{e-2d+c}, \ldots$$

From the second difference, one can extract a third difference, etc.

$$\underbrace{(d-2c+b)-(c-2b+a)}_{d-3c+3b-a}, \underbrace{(e-2d+c)-(d-2c+b)}_{e-3d+3c-b}, \ldots$$

Euler then constructs a series *equivalent* to A by setting $\alpha =$ to the first member of the first difference of A, $\beta =$ the first member of the second difference, etc. In this way he gets the following values: $\alpha = (b-a), \beta = (c-2b+a), \gamma = (d-3c+3b-a)$, etc. The equivalent series is given below.[8]

$$A = \frac{a}{2} - \frac{\alpha}{4} + \frac{\beta}{8} - \frac{\gamma}{16} + \frac{\delta}{32} - \cdots$$

Since the resulting series alternates as well, the same method can be reapplied to it. Using this method, Euler returns to the series w. By several iterations of the method of differences, along with the 'telescoping'[9] of terms, he obtains a result.

$$w = \frac{5}{16} + \frac{512}{2048} + \frac{2046}{131072} + \cdots$$

This makes $w = 0.581$ (Euler [1760]: 216). Yet, this is only the first approximation of the series value. Continued calculation and manipulation gives Euler the value: $w = 0.05963479$. Sandifer explains Euler's dialectical or rhetorical position in terms of the debate over the use of divergent series.

---

[8]Sandifer ([2006]: 4) "proves" this result on the basis of "an only slightly obscure identity about binomial coefficients":

$$\sum_{k=n}^{\infty} \frac{1}{2^{k}+1} \binom{k}{n} = 1$$

[9]A sum *telescopes* completely if its intermediate terms cancel and leave only the first and last terms of the series. In this case, the telescoping is incomplete and leads simply to a reduction of some of the terms in the series. It might better be termed *neglecting* terms.

> This is the key result of this paper, but Euler understands that some readers might not be convinced that he hasn't made any mistakes. So, he solves the same problem several other ways. For example, he finds diverging series for $\frac{1}{w}$ and $\log w$, and finds that similar methods also lead to a value of w near 0.59. He finds a way to write w and $\frac{1}{w}$ as continued fractions and evaluates those continued fractions to get still more estimates consistent with the ones before.[10]
>
> (Sandifer [2006]: 7)

Sandifer concludes the article by suggesting the rationality of Euler's procedure.

> By the end of the article, Euler has estimated w at least six different ways, and every time he gets the same estimate. *When such different analyses all lead to the same conclusion, it is easy to understand why mathematicians of Euler's time believed in the utility of [divergent series]*.
>
> (Sandifer [2006]: 8, emphasis added)

The method is clearly based on *extrinsic justification*. And it seems rational even if our confidence in the result cannot flow through the conduit of deduction. The justification is extrinsic because it finds its corroborating evidence outside the original methods. This is similar to using a variety of sieve methods to determine whether a given number is prime. Since each sieve has a possible error, none is justified on its own. But, as more sieves return the same result, confidence in the result rises even if it is never complete. Except for his explicit rejection of inductivism, one would probably think that this result accords nicely with Lakatos's methodology. Lakatos, following Popper, championed knowing fallibly over knowing with certainty.

3.2 LAPLACE'S LUCK

As noted above, the most intriguing part of Gödel's claims regarding extrinsic justification involve success or fruitfulness in *physics* as justification for the mathematics employed. Such justification is doubly

---

[10]Sandifer, following Euler, uses A where I have used w.

extrinsic in that it is both outside the mathematical methodology and outside of mathematics proper.

One way to understand such justifications for mathematics is to see them as related to naturalism, confirmational holism, and the so-called Indispensability Arguments. These arguments involve the ontological commitments of mathematics in virtue of the indispensability of mathematics for our best physical theories. Put roughly, this means that as an element of physical theory, math shares in the spoils of any confirmation (or refutation) of the physical theory.

A remarkable example of extrinsic justification through physical application is Laplace's estimation of the secular perturbation of the Earth's orbit around the sun due to Jupiter. This is a 3-body problem. As an instance of the more general *n-body* problem, keep in mind that there are linear solutions to 2-body instances. But, when a third body is added to the mix, the problem becomes much more difficult because when an orbital system contains (at least) three bodies, no exact solutions obtain for the set of equations that describe the orbital system. Still, an approximate solution obtains by expanding the series contained within the descriptive equations. This is known as the method of perturbations. It is so-called because cases in which the motion of two such bodies whose movements require only their mutual gravitational attraction will move along paths expressed by conic sections. Such motions are said to be unperturbed. But, when the action (and gravitational attraction) of a third body enter the system, or when the mass is distributed in non-spherical ways, or when there is resistance from a medium, the motion of the bodies is perturbed so that they follow non-conic paths. The perturbation methods allow for the approximation of the influence of these perturbing elements.

Laplace employed perturbation methods to solve problems in what is sometimes called *positional astronomy*, i.e., predicting/determining the position of astronomical objects on basis of known orbits, sizes and the like. In his compendium of celestial mechanics, Laplace worked to develop approximations using perturbation methods.[11] On this basis, for example, he was able to determine the effect Jupiter has on the

---

[11] Chapters 3 and 4 of the second book of the *Traité de méchanique céleste* give the formal elements of Laplace's investigations into perturbation methods.

Earth's orbit. Yet, the series through which Laplace was able to make these predictions were divergent.

> Laplace seems to have regarded analysis merely as a means of attacking physical problems, though the ability with which he invented the necessary analysis is almost phenomenal. As long as his results were true he took but little trouble to explain the steps by which he arrived at them; he never studied elegance or symmetry in his processes, and it was sufficient for him if he could by any means solve the particular question he was discussing. (Rouse Ball [1908]: 420)

Because the results of the approximation were so accurate, Laplace, and others, were confident in the correctness of the method. Yet, the proof that such methods were reliable wouldn't come for another hundred years.[12] The reason that the use of divergent series works in these cases is that the series are what we would now call *asymptotic series* or *semi-convergent series*. Such series seem to converge over a given interval, and diverge outside that interval. Euler's method of manipulating alternating series as detailed in 3.1 above, for example, forces the series to *converge* more quickly over its interval of convergence. So, the error measure of the approximation can be made arbitrarily small. This guarantees the utility of the method.

Still, one needs to keep in mind the seriousness with which Laplace depended upon observation and application as justification for the mathematical theory. The first volume of Laplace's major work on celestial mechanics dates from 1799. But already in 1770, he was regularly in the practice of verifying mathematical calculations with experimental data.

> It is interesting to note that not only does Laplace's interest in the mechanics of the solar system date from 1770, but so does his life-long habit of comparing the results of theory with observation. (Stigler [1978]: 238)

Was Laplace simply lucky that every questionable manipulation employed an *asymptotic series*? If Laplace employed *mathematical inductivism*, and there is no deductive route by which truth flows upwards,

---

[12]Poincaré and Stieltjes, working independently defined these series in order to explain their utility. See Kline [1972]: 1104ff.

then he was merely epistemologically lucky. However, it may not be as easy to reject the upwards flow of truth.

4 Quasi-Empiricism Reconsidered

The simplified account of Lakatosian theory types tracks them across two dimensions: the kind of value moving within a purely deductive system and the direction or current the value travels as it flows within the system. Deduction is the only mechanism or conduit through which the value flows on this account. Hence, insofar as there is no purely deductive mechanism by which truth travels from consequences to grounds, *inductivism*, or what would better be termed *mathematical abductivism*, the apparent *verification* of the grounds by the consequences must be left unexplained or rejected. Moreover, since Lakatos rejects Euclidean justification from above, the only truth-value flowing within the purely deductive system of mathematics is falsity. This explains why proofs can generate, at best, fallible knowledge.

To be sure, this explanation assumes that theories are systems of purely deductive relations. Deduction preserves truth from grounds to consequences and falsity from consequences to grounds. Yet, Lakatos wants to consider a broader range of possible values flowing within a system.

> We may speak, even more generally, of Euclidean versus quasi-empirical theories independently of *what* flows in the logical channels: certain or fallible truth and falsehood, probability and improbability, moral desirability or undesirability, etc. It is the *how* of the flow that is decisive.
>
> (Lakatos [1978a]: 29)

Lakatos focused his attention on directionality rather than value in part because he didn't limit the class of potential falsifiers to logical contradictions. This is a merely formal point regarding acceptable inferences. Let Δ_1 be a deductively organized theory. Let the grounds of the theory, in the form of axioms and postulates, be given by the set Γ_{Δ_1}. For any set of grounds, Γ_j such that $\Gamma_j \subset \Gamma_{\Delta_1}$ and $\Gamma_j \vdash \kappa_j \wedge \neg \kappa_j$, the contradiction, $\kappa_j \wedge \neg \kappa_j$ falsifies Δ_1.

Problems arise for this account, however, when the deductive system, Δ_1, formalizes some informal theory, call it $\mathbb{I}_1$. Let the grounds of $\mathbb{I}_1$ be the set Π, and let the consequences or theorems be any τ_i such that $\Pi \vdash \tau_i$. Now, suppose some $\tau_i = \neg \kappa_i$. Moreover, suppose that $\Gamma_{\Delta_1} \vdash \kappa_i$, but $\Gamma_{\Delta_1} \nvdash \neg \kappa_i$. As a formalization of $\mathbb{I}_1$, something is wrong with Δ_1; but is Δ_1 false?

> If we insist that a formal theory should be the formalization of some informal theory, then a formal theory may be said to be 'refuted' if one of its theorems is negated by the corresponding theorem of the informal theory.
> (Lakatos [1978a]: 36)

The problem, of course, is that one cannot *deduce* the falsity of any axiom or postulate of the formal theory Δ_1 on the basis of $\neg \kappa$. Lakatos is well aware of this. Hence, he terms this refutation *heuristic falsification*.

> A heuristic falsifier after all is a falsifier in a Pickwickian sense: It does not falsify the hypothesis, it only suggests a falsification—and suggestions can be ignored.
> (Lakatos [1978a]: 40)

A heuristic falsifier is neither injected into a deductive system nor does it flow through logical channels within a deductive system. Rather, heuristic falsification is an inter-theoretic judgment regarding the unacceptability of a formal theory as a representation/formalization of some underlying informal theory. Logical falsification, on the other hand, is intra-theoretic. Logical falsification reveals an inconsistency within the formal theory.

Now, why not include Laplace's inductivism in Lakatos's heuristic toolbox? One thing initially undermines this suggestion: Lakatos rejects inductivism in any form—he calls it delusional and one of the 'gravest' dangers in contemporary philosophy of mathematics (Lakatos [1978a]: 42). Of course, it would be anti-Lakatosian to allow some method for which truth infallibly inundates a system. But this isn't what inductivism offers. Indeed, so long as inductivism is understood in terms of (extrinsic) *evidence*,[13] any knowledge based upon an inductivist methodology will be fallible.

Two questions remain. First, is inductivism a mathematical methodology? And, if so, is it exhausted by what would normally be considered *application*? I expect that Lakatos did not think that inductivism, even as an heuristic method, was a part of mathematics proper. There is nothing like it in, for example, *Proofs and Refutations* (Lakatos [1976]). The only heuristic methods he considers in mathematics are comparisons of an informal theory with its proposed formalization. An inductivist methodology is quite different from this. Part of the difficulty is that whereas an heuristic falsifier cannot be injected into a formal system of mathematics, extrinsic evidence needn't be meta-theoretical in this same way. For Laplace a successful application injected truth into the mathematical system. Once truth gets into the system at the bottom, one cannot explain its upward movement in terms of deduction. Yet, if we allow for non-deductive channels within a system of mathematics, then there can be little resistance to any value moving in any direction within the system. Put another way, if truth is allowed to inundate a system from below through non deductive channels, then why not allow falsity to inundate a system from above through non-deductive channels? The simplest response is that Lakatos wouldn't. The only consequence relation he envisions is deductive. Hence, even if he were to allow for extrinsic evidence, it wouldn't have any justificatory force because there wouldn't be any conduit or channel through which it can flow through the system. This is why heuristic falsifiers are falsifiers in name only. They don't enter a system of deductive relations to determine the literal falsity of the grounds of the system. Rather, they show that the formalized deductive relations of one theory fail to capture the informal content of another theory. The formalized deductive relations of the *refuted* theory, are not more likely to turn out false, however, on the basis of heuristic refutation. Yet, the opposite seems to be exactly what extrinsic justification supports. The more extrinsic justification one can amass for a set of grounds, the more likely it is *to be true*. This

---

[13] Here the contrast is between *evidential* justification and demonstrative justification. Whereas the latter gives final or conclusive justification to its consequences, the former supports its consequence without demonstrating them conclusively.

means that a mathematical quasi-empiricism which allows for extrinsic evidence would be quite different from Lakatos's quasi-empiricism.

Finally, in answer to the second question, successful application is probably only one of many acceptable inductivist maneuvers. Maddy ([1988a] and [1988b]) offers an interesting starting point. She lists ten different kinds of extrinsic evidence in set theory (Maddy [1988b]: 758–9). Yet, only the last one is explicitly inter-theoretical. The others needn't be inter-theoretical. Some, for example, providing proofs to prior conjectures, may operate wholly within an informal framework. As such, they don't fit Lakatos's model for heuristic methods. Still, one would expect an account of mathematical justificatory practice to explain acceptability or unacceptability of such methods. A broader quasi-empiricism that allows for extrinsic justification as well as heuristic falsification is Lakatosian in spirit. And this suggests that Gödel was right to say that some axiom may be so fruitful that we must accept it as true.

Bibliography

Abel, N.H. [1964]. *Œvres complètes*. Ed. by S. Lie and L. Sylow. Vol. 2. First published in 1881. Christiana.

Euler, L. [1760]. "De seriebus divergentibus." *Novi Commentarii academiae scientiarum Petropolitanae*. Vol. 5. 205–37.

Gillispie, C.S. [1997]. *Pierre-Simon Laplace, 1748–1827: A life in exact science*. Princeton: Princeton University Press.

Gödel, K. [1947]. "What is Cantor's Continuum Hypothesis?" *American Mathematical Monthly* **54**. 515–25.

— [1983]. "What is Cantor's Continuum Hypothesis?" *Philosophy of Mathematics: Selected Readings*. Ed. by Paul Benacerraf and Hilary Putnam. First published in 1964. Revised. Cambridge: Cambridge University Press. 470–85.

Grattan-Guinness, I. [1990]. "Thus it mysteriously appears: Impressions of Laplace's use of series." *Rechnen mit dem Unedlichen: Beiträge zur Entwicklung eines kontroversen Gegenstandes*. Ed. by Detlef Spalt. Basel. 96–102.

Kline, M. [1972]. *Mathematical Thought from Ancient to Modern Times.* Oxford: Oxford University Press.

Lakatos, I. [1976]. *Proofs and Refutations.* Cambridge: Cambridge University Press.

— [1978a]. "A renaissance of empiricism in philosophy of mathematics." *Mathematics, science and epistemology.* Ed. by John Worrall and G. Currie. Cambridge: Cambridge University Press. 24–42.

— [1978b]. "Infinite regress and foundations of mathmatics." *Mathematics, science and epistemology.* Ed. by John Worrall and G. Currie. Cambridge: Cambridge University Press. 3–23.

— [1978c]. *Mathematics, science and epistemology.* Ed. by John Worrall and G. Currie. Cambridge: Cambridge University Press.

Laplace, S. [1825]. *Traité du méchanique céleste.* First published in 1789. Gauthier-Villars.

— [1904]. *Œuvres complète.* First published in 1891. Gauthier-Villars.

Maddy, P. [1988a]. "Believing the Axioms, I." *Journal of Symbolic Logic* **53** (2). 481–511.

— [1988b]. "Believing the Axioms, II." *Journal of Symbolic Logic* **53** (3). 736–64.

— [1992]. *Realism in Mathematics.* Oxford: Oxford University Press.

Popper, K. [2002]. *Conjectures and Refutations.* First published in 1963. London: Routledge and Keegan Paul.

Rouse Ball, W. [1908]. *A short history of mathematics.* New York: Dover.

Sandifer, E. [2006]. "Divergent series." *How Euler Did It.* MAA Online.

Stigler, S. [1978]. "Laplace's Early Work: Chronology and Citations." *Isis* **69** (2). 234–54.

ESSAY 4

Mathematical Concepts in Computer Proofs

Koen Vervloesem

The twentieth century has seen an important revolution in mathematical practice. For the first time in history, computers can take part in finding proofs. In 1969 Elsie Cerutti and Philip Davis experimented with a computer program proving Pappus' theorem (Cerutti and Davis [1969]). In 1976 Kenneth Appel, Wolfgang Haken, and their student John Koch proved the four-colour theorem, thanks to extensive computations (Appel and Haken [1977a] and Appel, Haken, and Koch [1977]). In 1989 Clement Lam and his colleagues verified with an extensive computer search that there do not exist finite planes of order 10 (Lam, Thiel, and Swiercz [1989]). In 1996 William McCune's program EQP proved the Robbins conjecture after eight days of computing (McCune [1997]). In 1998 Thomas Hales and his PhD student Samuel Ferguson proved Kepler's conjecture. Their proof relies on extensive computations (Hales [2005]).

These are not the only known computer proofs.[1] A lot of lesser known computer proofs exist in the wild. However, if one looks at mathematical practice, it seems that most mathematicians do not use computers in their activity of proving theorems. That is, computer algebra software like Mathematica, Maple, and Matlab are in the standard toolbox of a mathematician, whereas computer provers are not. The preceding list of success stories shows that computers *can* help mathematicians to prove theorems that are otherwise too difficult. So why do mathematicians not make use of them more often? We will look at this problem from the view of mathematical practice. According to computer scientist Ursula Martin mathematical practice can be defined as:

> [P]roducing conjectural mathematical knowledge by means of speculation, heuristic arguments, examples and experiments, which may then be confirmed as theorems by producing proofs in accordance with a community standard of rigour, which may be read by the community in a variety of ways. (Martin [1999])

Computer proofs can have a place in the second part of Martin's definition.[2] Hence, we can ask if a proof constructed by a computer corresponds to the community standard of rigour of mathematicians. One aspect of this standard is reliability, but another one is *understandability*. In this paper, I focus on the latter aspect. I start with

---

[1] In this article, when I talk about a 'computer proof,' this can be one of three types of proofs: (i) A *computer-assisted proof*, which is a proof of which a part has been verified by computer calculations. An example is the original proof of the four-colour theorem; (ii) a proof found by an *automated theorem prover*, by reasoning which steps have to be taken to prove the theorem. An example is EQP's proof of the Robbins conjecture; and (iii) a *formal verification*, which is a complete verification of a proof by a computer, entered by a human mathematician in an appropriate format. An example is Gonthier's formal verification of the four-colour theorem in Coq.

[2] Actually, in both parts. In this paper, I focus on the proving part. For the conjectural production, see Simon Colton's computer program HR, Colton, Bundy, and Walsh [1999], Colton, Bundy, and Walsh [2000].

the observation that there is a difference between understanding a computer proof and a human proof. To explain this difference, we have to look at some properties of proofs, such as the structure of the proof, the size of the proof, and the concepts used. In this paper I claim that an important difference is to be found in the concepts used in computer proofs and human proofs.

First this claim will be illustrated in its broader context with a typical example of a simple theorem and two different proofs of it. Most mathematicians will prefer the more conceptual one. Then some cases of computer proofs and the mathematical concepts they use are analyzed. They score poorly on the conceptual level. I will then use these observations to explain two things: (i) why many mathematicians object to computer proofs because of a lack of understanding and (ii) why computers cannot prove certain simple results.

1 Mathematical Concepts and Proofs

When a mathematician tries to prove a conjecture, he does not do this by beginning from the axioms and then trying to deduce the conjecture. Proof, conjecture, and conceptualisation are inextricably linked in the activity of mathematicians. David Corfield describes this process like this:

> Mathematicians perform these activities simultaneously—while clarifying a concept they notice a property which looks like it may hold for all of some class of objects, and while trying to prove that this is so, they find that it pays to introduce conceptual distinctions between elements of that class. (Corfield [2003]: 35)

In this paper I use the term *(mathematical) concept* to denote an abstract idea in mathematics. Examples are *number, matrix, set, group, norm,* and *cardinality*. We have an intuitive grasp of these mathematical concepts, particularly of 'basic' concepts like *set*, and *number*, which we make use of in our daily life. We can axiomatize a mathematical concept by 'unfolding' the concept and describing it in a compact and systematic way. Antony Eagle gives the following definition of an axiomatization of a concept:

> $A_1, \ldots, A_n$ are *axioms* (or axiom schemata) for a concept $\mathcal{M}$ iff $\{A_1, \ldots, A_n\}$ are a simple and concise set of independent claims (or claim schemata) that entail the core platitudes about $\mathcal{M}$, and do not entail the negation of any core platitude about $\mathcal{M}$.[3] (Eagle [2008]: 78)

Kurt Gödel is famous for his emphasis on (conceptual) intuitions as a prerequisite for mathematical knowledge. Eagle quotes Gödel saying:

> [M]athematical concepts must be sufficiently clear for us to be able to recognize their soundness and the truth of the axioms concerning them[.][4] (Eagle [2008]: 69)

In this section I describe a fairly simple theorem and three different proofs of it. The proofs differ in the mathematical concepts they use. The theorem is Cassini's identity for Fibonacci numbers.

We first have to define the Fibonacci numbers.

DEFINITION: The Fibonacci numbers form a sequence defined by the following recurrence relation:

- $F_0 = 0$.
- $F_1 = 1$.
- $F_n = F_{n-1} + F_{n-2}$ for $n > 1$.

There are a lot of results known for this sequence, such as this curious formula:

THEOREM 1: *Cassini's identity* $F_{n+1}F_{n-1} - F_n^2 = (-1)^n$ for $n \geq 1$.

The following three proofs can be found in the literature.

---

[3] Eagle comments that this axiomatization is "not *trivial*, because it can be quite difficult to decide which set of claims is genuinely simple, and genuinely captures all of the platitudes about the concept in question."

[4] The quote is from Kurt Gödel, 'What is Cantor's continuum problem?' in Paul Benacerraf and Hilary Putnam (eds.), *Philosophy of mathematics: Selected readings*: 470–85 (1964).

1.1 PROOF BY INDUCTION

PROOF: We prove by induction that the identity holds for all $n \geq 1$. If $n = 1$, the numbers F_{n-1}, F_n, and F_{n+1} become F_0, F_1 and F_2, respectively. The identity becomes $1 \cdot 0 - 1^2 = (-1)^1$, which is correct. Assuming that the identity holds for $n = m$, with $m \geq 1$, we check that the identity holds for $m+1$: $F_{m+2}F_m - F_{m+1}^2 = F_{m+1}(F_m - F_{m+1}) + F_m^2 = -F_{m+1}F_{m-1} + F_m^2 = -(-1)^m = (-1)^{m+1}$. By our induction hypothesis, the identity holds now for $n = m+1$. This concludes the proof.[5] □

1.2 PROOF BY BINET'S FORMULA

We can obtain a conceptually more advanced proof if we use Binet's formula for the Fibonacci numbers. Binet's formula is a closed formula for the recursively defined Fibonacci numbers: $F_n = \frac{\phi^n - (1-\phi)^n}{\sqrt{5}}$. The number ϕ in this formula is the so-called *golden ratio*, equal to $\frac{1+\sqrt{5}}{2}$ or approximately 1.618. It is one of the solutions of the quadratic equation $x^2 - x - 1 = 0$. Binet's formula can be proved easily by induction.[6] Using this formula, one can obtain the following proof of Cassini's identity.

PROOF: Substituting Binet's formula in the left-hand side of Cassini's identity and simplifying the formula, we obtain $(\phi - \phi^2)^n$. The expression $\phi - \phi^2 = -1$, because ϕ is a solution of the quadratic equation $x^2 - x - 1 = 0$. Hence, $F_{n+1}F_{n-1} - F_n^2 = (-1)^n$ for $n \geq 1$. □

Petkovšek, Wilf, and Zeilberger ([1996]: 12) note that many identities between Fibonacci numbers can be proved easily by using Binet's formula. They even give an example of 'proving' Cassini's identity in the computer algebra package Maple.

---

[5]Thanks to Herman Roelants, who simplified my original proof considerably.

[6]A proof sketch goes roughly like this: Calculate and simplify the expressions for the successive powers of ϕ, using the relation $\phi^2 = \phi + 1$. The list begins as follows: $\phi = 1\phi + 0$, $\phi^2 = 1\phi + 1$, $\phi^3 = 2\phi + 1$, $\phi^4 = 3\phi + 2$, $\phi^5 = 5\phi + 3$, $\phi^6 = 8\phi + 5, \ldots$ The coefficients of these equations are all consecutive Fibonacci numbers. By induction, we can prove the generalized equation $\phi^n = F_n\phi + F_{n-1}$. Because $1 - \phi$ is also a solution of $x^2 - x - 1 = 0$, we have $\phi^n = F_n\phi + F_{n-1}$ and $(1-\phi)^n = F_n(1-\phi) + F_{n-1}$. Subtracting the second equation from the first, we obtain: $\phi^n - (1-\phi)^n = F_n(2\phi - 1)$. Solving for F_n and simplifying, we obtain Binet's formula: $F_n = \frac{\phi^n - (1-\phi)^n}{\sqrt{5}}$. □

1.3 PROOF BY MATRIX THEORY

A key insight for another proof is that the left-hand side of Cassini's identity can be interpreted as the determinant of a 2 by 2 matrix of Fibonacci numbers, $\begin{bmatrix} F_{n+1} & F_n \\ F_n & F_{n-1} \end{bmatrix}$. Another well-known theorem[7] decomposes this matrix as the n-th power of a matrix with determinant -1, which gives us the following succinct argument.

PROOF:

$$F_{n-1}F_{n+1} - F_n^2 = \det \begin{bmatrix} F_{n+1} & F_n \\ F_n & F_{n-1} \end{bmatrix}$$

$$= \det \begin{bmatrix} 1 & 1 \\ 1 & 0 \end{bmatrix}^n = \left(\det \begin{bmatrix} 1 & 1 \\ 1 & 0 \end{bmatrix} \right)^n = (-1)^n. \quad \square$$

1.4 THE CONCEPTS IN THE THREE PROOFS

If we look at these three proofs, we see a clear difference in proof strategy. The first proof uses a fairly limited set of concepts: the definition of Fibonacci numbers and some algebraic calculations. Induction does the rest. It is the kind of proof that a high-school student interested in Fibonacci numbers would fabricate. You do not need insight to find such a proof, just some fiddling with formulas. Reading this proof, one does not get the idea *why* Cassini's identity holds. For example, one cannot explain in one sentence why it holds. The calculation looks nice, but you cannot see an idea, however short the calculation is.

The second proof already uses some more concepts. We use another formula for the Fibonacci numbers, but the rest is just an ordinary calculation. It is a routine proof, as the example in Maple shows. Even here, one does not get an idea why the identity holds: It cannot be explained in one sentence.

The third proof is completely different and it is based on a lot more mathematical concepts. Next to the definition of Fibonacci numbers, one uses the matrix of Fibonacci numbers in the form of the n-th power

---

[7] If we define the matrix $Q = \begin{bmatrix} 1 & 1 \\ 1 & 0 \end{bmatrix}$ (the so-called *Fibonacci Q-matrix*), then one can prove by induction that $Q^n = \begin{bmatrix} F_{n+1} & F_n \\ F_n & F_{n-1} \end{bmatrix}$.

of the matrix $\left[\begin{smallmatrix}1&1\\1&0\end{smallmatrix}\right]$. By shifting the level of concepts from the domain of integers to the domain of matrices, we got access to the powerful determinant concept. This makes it possible to consider the left-hand side of the identity as a determinant of a matrix of Fibonacci numbers. After this key insight, the rest of the proof follows almost trivially.[8] The calculations in the proof are trivial and one uses some basic properties. Eventually, this proof shows why the identity holds, e.g. one sees why the −1 and the n-th power occur on the right-hand side of the identity. A one-sentence explanation of this proof could be: The left-hand side of Cassini's identity can be rewritten as the n-th power of a determinant with value −1 of a matrix of Fibonacci numbers.[9]

Why the analysis of this example?[10] Many computer proofs have access to a limited set of concepts which they use to calculate. My claim is that these proofs look a lot like our first two proofs by calculation of Cassini's identity and that people do not like them because they give no insight.

1.5 WHICH PROOFS GIVE INSIGHT?

One of the research domains in philosophy of mathematical practice is *mathematical explanation*, with questions like "Which mathematical proofs give insight and explain why a theorem holds?" One account of this problem is given by Michael Resnik and David Kushner, following

---

[8] As Kerber and Pollet ([2002]: 9) say, "Sometimes an appropriate reformulation of a problem into another representation is already the key step to find a proof. Different representations allow to apply knowledge from different sources to a problem."

[9] Some people told me that I am cheating in this example. Their reasoning is something like this: "Yes, this proof is simple, but that's because you are using the 'well-known theorem' as a lemma in a vital step of the proof." I agree, but that's what mathematicians do: They are always 'cheating' by using lemmas, proof schemas, etc.

[10] Jeremy Avigad ([2006]) gives an analysis of other examples of different proofs of a theorem and their virtues, e.g. the theorem that $2^{2^5}+1$ is not a Fermat prime and the theorem that xy can be written as a sum of two squares if the natural numbers x and y can. But while I focus on the *concepts* in my comparison of different proofs, Avigad focuses on *generalizability* and other criteria. However, both discussions are related. In the example of sums of squares, Avigad notes that introducing the concept *norm* makes a proof with Gauss integers generalizable: If we replace the complex numbers by quaternions or octonions and keep using the formulation of the theorem with norms, we get product rules for sums of four or eight squares, respectively.

Bas van Fraassen who says (in the context of *scientific* explanations) that an explanation is always context-dependent. One cannot say that a proof gives insight "tout court":

> Whether or not something is evident from a proof is relative to subgroups of the mathematical community, at best. A proof that explains to a mathematical logician may be anything but evident to a topologist.
> (Resnik and Kushner [1987]: 146)

We can also see this in our example. A high-school student will find his elementary proof by induction more explanatory than the proof by matrix theory, while a professional mathematician will consider the explanation of the last proof better.[11] It all depends on the background of the person trying to understand the proof.[12]

According to van Fraassen, asking for a (scientific) explanation comes down to asking a *why-question*.[13] Resnik and Kushner apply this to mathematical explanations. Whether a proof explains or not depends on the why-question we ask. The authors remark that so-called "explanatory proofs" are then proofs giving answers to more why-questions.[14]

If we apply the why-questions approach of Resnik and Kushner to our three proofs of Cassini's identity, then we see why most mathematicians prefer the third proof. If we simply want to know why the result is true, rather than false, then all three proofs suffice. However, if we want to know why the right-hand side of the identity has an exponent n or why

---

[11] A mathematician will consider this explanation also *more pleasing*, but that's an aesthetic evaluation, not a cognitive one. Unless, of course, one is willing to argue that both are connected.

[12] Thanks to Dirk Schlimm, who made the importance of this point clear to me.

[13] Van Fraassen [1980]: 134, "An explanation is an answer to a why-question. So, a theory of explanation must be a theory of why-questions."

[14] Resnik and Kushner [1987]: 154, "Now the so-called explanatory proofs, the ones which 'reveal the heart of the matter,' present more information and do so more persicuously than do 'nonexplanatory' proofs of the same results. Thus they provide the ingredients for answering more why-questions than other proofs. But they are not explanatory in and of themselves." However, David Sandborg in his [1998] maintains that the why-question approach misses crucial aspects of certain explanatory evaluations. He shows that explanatory judgements can be context-dependent without being relative to why-questions.

the number −1 occurs, then the third proof gives a more perspicuous answer. This does not mean that one can always find a proof which is 'the best' proof of a theorem. Different proofs can enlighten different aspects of the theorem.[15]

2 Mathematical Concepts in Computer Proofs

In this section I analyze some examples of computer proofs, the concepts they use, and the insights they give us.

2.1 THE FOUR-COLOUR THEOREM

THEOREM 2: *The four-colour theorem* Any planar map can be coloured using four colours in such a way that regions sharing a common boundary, other than a single point, do not have the same color.

There exist several proofs of this theorem: Appel, Haken and Koch, (1977)[16], Robertson, Sanders, Seymour, and Thomas (1996)[17], and Gonthier and Werner (2004)[18]. All proofs have the same high-level structure of finding an unavoidable set of irreducible configurations, ultimately based on Kempe's first incorrect proof (Kempe [1879]).[19] One can clearly see an evolution in these proofs from a computer-assisted proof to a completely computer-verified proof. The proof by Appel and Haken uses an unavoidable set of 1482 configurations, and 487 'discharging rules.' Robertson and his colleagues simplified this to 633 configu-

---

[15] Even the authors of *Proofs from The Book*, which is supposed to be a selection of 'the perfect proofs for mathematical theorems,' regularly give a couple of proofs for a certain theorem. For example, in chapter 7, *Three times* $\pi^2/6$ (Aigner and Ziegler [2004]: 35), they give three proofs of Euler's famous identity $\sum_{n\geq 1}(\frac{1}{n^2}) = \frac{\pi^2}{6}$. One of the proofs answers how fast (or rather how slow) the series converges to $\frac{\pi^2}{6}$, while the other proofs do not answer this question. However, the other two proofs give more insight in the geometric interpretation of the identity. So there is not one unique best proof of the identity.

[16] Appel and Haken [1977a] and Appel, Haken, and Koch [1977]

[17] Robertson et al. ([1997])

[18] Gonthier ([2004])

[19] A good exposition of the techniques in the proofs can be found in Appel and Haken [1977b] and [1978].

rations and 32 discharging rules. Gonthier used the configurations and rules of Robertson's proof and formalized this completely in Coq.

A problem that occurs in many computer proofs in geometry is that the mathematician has to translate geometrical concepts to algebraic concepts, before feeding them to the computer. While mathematicians have no problems reasoning with geometrical concepts like point, line, etc., because they can use their geometrical intuitions of everyday life, computers do not have this advantage. The whole reasoning of the computer is algebraic, which makes the proof more difficult to understand by a mathematician, because in the computerized proof there are virtually no references to the original geometrical concepts the theorem is about. The mathematician can verify the algebraic derivations, but he cannot use his geometrical intuitions to understand them.

An example that shows this clearly is Gonthier's complete formalization in Coq of Robertson's proof of the four-colour theorem. Gonthier could not use the 'intuitive' concepts in Robertson's proof for a completely formalized proof, because:

> [M]ost graph theory proofs rely on the visual analysis faculties of the reader, and a proof assistant like COQ is fully devoid of such faculties, we mostly had to come up with our own proofs (sometimes using visual intuition as guidance, but not always). (Gonthier [2004]: 17)

So Gonthier had to formalize these concepts in the form of combinatorial properties, which can be manipulated perfectly by a computer program. He tried different approaches to define the concept of a plane graph. About one approach using the Jordan curve theorem, he writes:

> However, this approach results in proofs that, while convincing for humans, are difficult to formalize because they contain an informal mix of combinatorics and topology.
> (Gonthier [2004]: 5)

Gonthier eventually used another approach, more optimized for formalizability than for intuitiveness. For this purpose, he used the concept of a *hypermap*. This made it possible for him to define the four-colour theorem in purely combinatorial terms, without making use of the Jordan curve theorem. Gonthier proved an equivalent of the Jordan curve theorem for hypermaps. This approach simplified parts of Robertson's proof. For example, while Robertson's proof had to refer to a *folklore theorem* with no known published proof, Gonthier could dispense with this theorem (Gonthier [2004]: 17). So hypermaps made the proof conceptually less intuitive, but structurally simpler.

In his discussion of his proof, Gonthier refers several times to the "gap between the intuitive, picture-rich proof outline, and the very precise logical statement that had to be fed to the Coq proof assistant." Concepts that we humans commonly use in a proof of a theorem in geometry are generally difficult to formalize. Gonthier mentions theorems like "[T]he whole map is planar if and only if the two pieces are." We can intuitively grasp such a theorem, but it turns out to be hard to formalize and even harder to understand the formalized version.

Gonthier considers his proof as a logical evolution from the proofs he built upon, towards more precision and more confidence in the correctness of the proof: "Our actual work was to find a systematic way of getting the details right—all of them." His proof is based upon Robertson's proof, which is in essence a simplification of the proof by Appel and Haken, which in turn uses the same basic outline as Kempe's incorrect proof. However, by getting all the details right, the proof has become less insightful. While his proof gives us finally enough confidence in the correctness of Robertson's proof, it does not give us more insight. For computer scientist Ursula Martin, such efforts for formalization are at odds with mathematical practice. The mathematician William Thurston agrees and he considers formalized mathematical proofs by computers a "worthwhile project," but warns us not to depend too much on it:

> We should recognize that the humanly understandable and humanly checkable proofs that we actually do are what is more important to us, and that they are quite different from formal proofs. (Thurston [1994]: 171)

So the mathematical concepts, and especially the formalized mathematical concepts, in Gonthier's proof are a problem for our understanding of the proof. Some mathematicians think the problem lies already in the combinatorial approach all these proofs are based upon. For example, Paul Halmos thinks we are on the wrong track for the four-colour theorem: The proofs are using the wrong concepts. He thinks that 100 years from now the four-colour theorem will be proved in a few pages in a first-year graduate course, thanks to the appropriate, but yet unknown, concepts (Halmos [1990]: 577).

2.2 THE ROBBINS PROBLEM

In 1933, E.V. Huntington proposed the three following axioms for Boolean algebra:

Commutativity $x + y = y + x$.

Associativity $(x + y) + z = x + (y + z)$.

Huntington axiom $\neg(\neg x + y) + \neg(\neg x + \neg y) = x$.

(Huntington [1933b] and [1933a])

We use + for the addition and ¬ for the complement here. If we begin from Huntington's axioms, we can prove that there exist constants 0 and 1 with their well-known properties,[20] so each Huntington algebra is a Boolean algebra. Herbert Robbins wondered if this still holds if he replaced the Huntington axiom by the following axiom:

$$\textit{Robbins axiom}: \neg(\neg(x + y) + \neg(x + \neg y)) = x$$

One can verify easily that the Robbins equation is valid in each Boolean algebra, so each Boolean algebra is a Robbins algebra. A more difficult question is: Do the commutativity, associativity, and Robbins axiom together form an axiomatization of Boolean algebra? Or, in short: Is each Robbins algebra a Boolean algebra? Robbins and Huntington could not find a proof, and neither could Tarski. A proof eventually was found by a computer program, EQP, designed by William McCune of the Argonne National Laboratory in Illinois.[21]

---

[20] $0 + x = x$ for all x in the Boolean algebra B and $\neg(\neg x + x) = 0$ for all x in B. The constant 1 has analog properties.

[21] EQP is an equational prover, using resolution and unification, as described in Alan Robinson's ([1965]) classical paper.

EQP's proof of the Robbins problem has other 'conceptual problems' than the proofs of the four-colour theorem. The Robbins problem uses already formal, algebraic concepts, so the problem is not that we have to convert intuitive concepts to formal concepts. The fact that EQP's proof gives us no insight why the theorem holds, is not entirely the computer's fault, but in part the fault of the problem itself.

EQP's proof consists of 12 steps with long formulas with many nested parentheses. So although the proof is short, it is difficult to understand. For example, one sentence of the proof reads as follows:

$$\neg(\neg(\neg(\neg(\neg x + x) + \neg(\neg x + x) + x + x + x + x) + \neg(\neg(\neg x + x) + x + x + x) + x) + x)$$
$$= \neg(\neg(\neg x + x) + \neg(\neg x + x) + x + x + x + x)$$

Mathematician Stanley Burris called such computer output unreadable. Moreover, each step in EQP's proof follows from other steps, but Burris had to fill in the details, such as which substitutions should be carried out (Burris [1996]). Burris reworked the short, condensed proof into a longer proof with simpler steps. As part of this, he introduced some auxiliary variables, such as $T := D + E + y + y$. His proof spans four pages: one page of definitions and auxiliary variables, and three pages for proof steps. This results in 105 short equations which can be checked easily, or as Burris says: "I ended up with a lot of little equations. You could easily sit on a bus and go through the hundred or so steps of the proof" (Peterson [1997]). He makes his point clear in the title of his manuscript: "An anthropomorphized version of McCune's machine proof that Robbins algebras are Boolean algebras."

Bernd Dahn analyzed EQP's proof with the computer program ILF, which simplified the formulas in the proof by introducing abbreviations for partial formulas that the proof used extensively (Dahn [1997]). A statistical analysis pointed out that the formula $\neg(\neg(\neg x + y) + z)$ occurred a couple of times. Introducing this formula in a definition did not simplify the proof as a whole, because the formula did not have useful properties. However, the simpler formula $\delta(x, y) = \neg(\neg x + y)$ did simplify the proof. This suggests that EQP's proof is not using the 'right' (or the most fruitful) concepts.

Branden Fitelson is on the same track: He suggests that EQP's proof is difficult to follow because it is not *conceptual*.[22] A human mathematician would never find a proof like EQP's because it is too complex.[23] The proof shows no vision, no strategy;[24] it is just the proof that the computer program accidentally found with a specific set of search parameters in a reasonable time. EQP supports equational reasoning in a deductive way, knows a limited set of symbolic concepts, and never introduces or invents new concepts. The result cannot be a conceptual, structured, abstract, or elegant proof.[25] Fitelson calls the unintuitiveness of these proofs a big problem:

> One of the most pressing current problems facing researchers in automated theorem-proving (according to Professor Kenneth Kunen, personal communication) is the translation and reconstruction of complex, unintuitive computer proofs into forms that are more readily understood by human beings.

---

[22] In Peterson [1997], he says: "It's a very complicated proof, not at all elegant or conceptual. It's a perfect example of the type of proof that a machine can find but we can't because of its complexity."

[23] Fitelson [1998], "The proof found by EQP is quite complex and difficult to follow. Some of the steps of the EQP proof require highly complex and unintuitive substitution strategies. As a result, it is nearly impossible to reconstruct or verify the computer proof of the Robbins conjecture entirely by hand."

[24] Of course the words *vision* and *strategy* are vague, but mathematicians do talk like this about proofs. The mathematician Alan Bundy has tried to formalize this sense of strategy in the notion of a *proof plan* that reduces the search space for an automated theorem prover and makes the proof easier to read for a human mathematician. In Bundy [1988] he describes this as follows: "We believe that human mathematicians can draw on an armoury of such proof plans when trying to prove theorems. It is our intuition that we do this when proving theorems, and the same intuition is reported by other experienced mathematicians. One can identify such proof plans by collecting similar proofs into families having a similar structure, e.g. those proved by *diagonalization* arguments. Many inductive proofs seem to have such a similar structure."

[25] One notable mathematician who disagrees is knot theorist Louis Kauffman, who said about the proof: "I understood EQP's proof with an enjoyment that was very much the same as the enjoyment that I get from a proof produced by a human being" (Kauffman [2001]). Other examples of more elegant and conceptual proofs found by automated theorem provers can be found in Wos and Pieper [2003] and Belinfante [1999].

3 The Omnipresence of Concepts

The importance of the 'right concepts' can be seen in definitions and in theorems using concepts from multiple domains. Both definitions and theorems with multiple concepts can be found in lots of mathematical texts. One cannot imagine mathematics without them. Hence, the problems computer provers have with concepts are not so light as it appears.

3.1 DEFINITIONS

Concepts regularly appear in *definitions*, an essential part of mathematics. An example is the epsilon–delta definition of continuity of real functions.

DEFINITION: A function f is continuous in x $\Leftrightarrow$ for each $\epsilon > 0$ there exists a $\delta > 0$ such that $|f(x) - f(y)| < \epsilon$ for all y with $|y - x| < \delta$.

If we grasp this definition and if we have proved some properties of the concept of continuity of a real function, then we can prove easily some other theorems about real functions. For example, if we have to prove the continuity of $f(x) = (x + 3)^{100}$, we can recognize that the function is a composition of two continuous functions. Referring to the theorem that the composition of two continuous functions is itself a continuous function, we have proved our result. Human mathematicians find proving such theorems by appealing to definitions very intuitive, but current computer provers have problems with this approach.[26]

Another important aspect of definitions of concepts is that they are needed in both directions in proofs. Beeson gives the example of the definition of a commutator in group theory (Beeson [2003]). A commutator $[x,y]$ is defined as $[x,y] = x^{-1}y^{-1}xy$. Even in fairly simple problems, computer provers have difficulty reasoning with the commutator concept. For example, in some situations we have to 'apply' the definition of a commutator from left to right: We substitute $[x,y]$ by $x^{-1}y^{-1}xy$, after which our expression becomes temporarily more difficult, with the possibility that we can simplify it by interaction with

---

[26] Beeson [2003], "Merely recognizing $f(x) = (x + 3)^{100}$ as a composition of two functions is beyond the reach of current theorem-provers—it is an application of the author's current research into 'second-order unification.'"

the symbols around the commutator expression. In other situations we have to apply the definition from right to left: We see a subexpression $x^{-1}y^{-1}xy$ in an expression and substitute it by $[x,y]$. This occurs when we have to prove that a certain difficult expression can be seen as a commutator. Computer provers cannot 'see' when they have to substitute the definition from left to right or vice versa.

3.2 FORMULATIONS WITH MULTIPLE CONCEPTS

Different types of concepts not only appear in proofs of theorems, but also in formulations of the theorems. A lot of theorems mention different mathematical concepts, with the consequence that proofs of them have to be conceptually rich. Beeson gives the example of theorems about regular n-polygons: These are theorems in plane geometry, but we also need the concept of a natural number n and we probably need a proof by induction for the theorems. These proofs in geometry will thus regularly mention concepts in the domain of natural numbers, next to geometrical concepts. And this is only first-order geometry, a 'simple' theory (Beeson [2003]).[27]

Another example of a domain where we need different concepts from various domains is ring theory. This is also a 'simple' theory, taught to university students in an introductory course of algebra. Beeson writes:

> In ring theory, one tries to prove a theorem using only the ring axioms; if one succeeds, the theorem will be true in all rings. However, in books on ring theory one finds many theorems about rings that are not formulated purely in the language of ring theory. These theorems have a larger context: they deal with rings and subrings, with homomorphisms and

---

[27] This does not rule out the possibility that in specific cases one can 'code' the natural numbers in geometric terms. Doing this, one can dispose of the natural numbers and keep using a pure geometrical theory. Using this coding, the formulation of the theorem of course becomes very complicated, so this is not really an improvement. This coding procedure can not be done in general: Alfred Tarski proved the completeness of his axiomatic formulation of first-order Euclidean geometry, while by Gödel's first incompleteness theorem Peano arithmetic is incomplete. Hence, arithmetic cannot be coded in first-order Euclidean geometry in a general way, if one needs full PA.

isomorphisms of rings, and with matrix rings.

(Beeson [2003])

This 'context' causes the appearance of concepts from other domains in ring theory. Homomorphisms and isomorphisms are functions from one ring to another one and functions are not mentioned in the axioms of ring theory. Matrix rings are rings which have matrices with coefficients of a given ring as its elements. This introduces the concept of a matrix in our theorems about ring theory. Moreover, some theorems are about matrices of arbitrary size, which silently introduces the concept of a natural number. In short, ring theory as an isolated theory is not of much use.

In his Master's thesis, Beeson's student Tony Huang tried to formalize 150 exercises of an algebra textbook in first-order ring theory (Huang [2002]). He could formalize only 14 exercises in first-order ring theory, while the other exercises were too complex because their formulation made already use of concepts from other domains. The automated theorem prover Otter could prove the first 14 easily, but the other ones were out of reach because of their formulations. Computer provers would have to learn to deal with concepts from different domains before they can prove these theorems.

Beeson mentions Lagrange's theorem as a typical theorem in group theory: If H is a subgroup of a finite group G, the number of elements of H is a divisor of the number of elements of G. Beeson does not expect a computer proof of Lagrange's theorem soon:

> At present, no theorem-proving program has ever generated a proof of Lagrange's theorem, even though the proof is very short and simple. The obstacle is the mingling of elements, subgroups, mappings, and natural numbers.[28]
>
> (Beeson [2003])

According to Beeson, only 10% of the problems in an introductory course of abstract algebra can be formulated in the first-order languages

---

[28]The proof is indeed "very short and simple": The cosets Hg form a partition of G. Since G is finite, there are a finite number of cosets, n. Since each coset has $|H|$ elements, we have $|G| = |H| \cdot n$, and so the number of elements of H is a divisor of the number of elements of G. □

of groups, rings, etc. As soon as the text introduces subgroups, homomorphisms, isomorphisms, and orders, concepts from different domains enter the scene. Applying first-order predicate logic to the first-order axioms of group theory will result in a lot of theorems, but scarcely any of the theorems in group theory textbooks or articles. Yehuda Rav concludes from this that we should not think about mathematicians as 'deduction machines,'[29] but as inventors of methods and concepts.[30]

4 Computation and Reasoning

Complex proofs generally consist of two types of steps: computations and reasoning. These two types have been part of mathematics since the Babylonians and Greeks of antiquity: Babylonian mathematicians excelled at calculations and algorithms, while they had no concept of rigorous proof. In contrast, Greek mathematicians invented the concept of a proof by reasoning, but Euclid had problems with algebraic equations, because when he needed them, he translated them to geometric propositions and tried to reason with them.[31]

Some simple theorems can be proved by appealing to computation or reasoning alone. We have already seen an example in the calculational proof of Cassini's identity by using Binet's formula. Even a computer algebra package like Maple can 'prove' this theorem just by calculating and simplifying the formula. Some other simple theorems can be proved by appealing only to some general reasoning principles, like mathematical induction or modus ponens.

If the theorems become more complex, however, you need computation *and* reasoning. The reasoning steps can account for the "proof plan" on a high level and for the connection of the different concepts used in the proof, which makes the proof more insightful, understand-

---

[29] Or, as Paul Erdős said, a machine turning coffee into theorems.

[30] Rav [1999]: 17, "This ought to stir the slumber of those who still think of mathematicians as deduction machines rather than creators of beautiful theories and inventors of methods and concepts to solve humanly meaningful problems."

[31] There is no sharp line between computation and reasoning. In a sense, it is a gradual distinction. However, in the rest of this paper I will take for granted that most proofs contain prototypical examples of computational and reasoning steps. It is these prototypical examples and their properties that are described in the rest of this paper.

able, and interesting for mathematicians. Michael Beeson describes the combination of computation and reasoning in proofs as follows:

> Typically computational steps move 'forwards' (from the known facts further facts are derived) and logical steps move 'backwards' (from the goal towards the hypothesis, as in *it would suffice to prove*). The mixture of logic and computation gives mathematics a rich structure that has not yet been captured, either in the formal systems of logic, or in computer programs. (Beeson [1998])

The computation steps account for the parts of the proof whose task is to verify some auxiliary steps in the proof. An example of this kind of computation can be found in an article by the mathematician David Mumford, who writes after the theorem that a certain function is correctly defined:

> The proof of this Proposition is a ghastly but wholly straightforward set of computations. It took me several hours to do every bit and as I was no wiser at the end—except that I knew the definition was correct—I shall omit details here. (Mumford [1967]: 230)

According to the Russian mathematician Yuri Manin, many mathematicians have such an experience with "mechanical proofs, even ones done by hand." His moral is: "[A] good proof makes us wiser" (Manin [1981]: 107). Mathematicians regularly accept a computation as a proof, but most of the time this is a verification of an auxiliary step or a result which is already understood well. Dominique Pastre, who designed the theorem prover Muscadet, writes:

> If a mathematician uses an automated theorem prover as an assistant and only wants to verify a minor point, a black box which answers a yes or no is sufficient, but if he wants to study a delicate point, it is necessary that the prover explains itself by giving important reasons and not tedious argumentation. (Pastre [1999]: 6)

4.1 CONCEPTS IN COMPUTATION AND REASONING

What have computation and reasoning to do with mathematical concepts? The difference is that concepts are manipulated in a different way in these two types of proof activities. A computation manipulates a *fairly limited* set of concepts, most of the time on the same level or in the same domain. Moreover, a computation goes (quasi-)automatically. At each point of the computation, you generally know what the next step is.

For example, a multiplication of two natural numbers only works in the domain of natural numbers. A calculation of a determinant of a matrix uses concepts from different domains: matrices and integers. The calculation is moving from the beginning to the end following a strict algorithm or some simple rules. One can essentially consider a computation as a black box: It does not matter what is happening inside, as long as you obtain the right output.

When one is using reasoning steps, one gets a different picture. In reasoning, one tries to relate some concepts, regularly from different domains. Moreover, reasoning steps require insight or trying many different possibilities. It is not an automatic process going forward and it is very difficult.

For example, our third proof of Cassini's identity, using the switch from number theory to matrix theory, is an example of this type of proof. Seeing that the left-hand side of Cassini's identity can be considered as a determinant of a matrix is not an automatic step. It requires insight, chance, or much trying, and knowledge of different mathematical domains. There is no (obvious) algorithm to carry out this step.

4.2 COMPUTATION AND REASONING IN COMPUTER PROOFS

On the one hand, proofs of the four-colour theorem and many other computer-assisted proofs make extensive use of computations. On the other hand, automated theorem provers like EQP or Otter make extensive use of reasoning, without computations for simplifying minor steps. So most of the time, computer provers are strong at computation and weak at reasoning, or vice versa. As we have seen before, a good proof of a complex theorem needs computation *and* reasoning, so most computer provers cannot provide us with good proofs. Beeson describes this situation as follows:

> What is interesting, and surprising to people outside the field, is that the mechanization of logic and the mechanization of computation have proceeded somewhat independently. We now have computer programs that can carry out very elaborate computations, and these programs are used by mathematicians 'as required.' We also have 'theorem-provers,' but for the most part, these two capabilities do not occur in the same program, and these programs do not even communicate usefully. (Beeson [2003])

Programs like Otter, based on the resolution method, are specialized in reasoning steps, but do not carry out much calculations. They have no *mathematical knowledge* and hence cannot execute calculations or substitutions that are simple for us. Human mathematicians can do this because by their experience they have built up a 'bag of tricks' with techniques to prove theorems of a certain form. When a mathematician wants to prove a new conjecture, he can use his reasoning powers, his calculation powers, and his bag of tricks of standard techniques. Beeson tried to mimic this approach in his computer program Weierstrass: This is a prover coupled to a database with mathematical knowledge and which tries to use a human approach to prove theorems (Beeson [1998]). One can say that Beeson injects mathematical knowledge into a theorem prover by giving it domain specific decision and heuristic procedures. This makes it possible for Weierstrass to reason and calculate with concepts in different domains.

Beeson did extensive experiments with his computer program Weierstrass combining computation and reasoning. The program automatically found epsilon–delta proofs of the continuity of certain functions, like x^n, $\sqrt{x}$, $\log x$, $\sin x$, and $\cos x$. The proofs consist of algebraic calculations and reasoning steps about inequalities. After some improvements Weierstrass even found automatically a proof of the irrationality of the number e.[32] The produced proof consists of various reasonings and cal-

---

[32] Beeson [2001], "A certain inequality involving factorials is needed in the course of the proof; the program finds a proof of this inequality by mathematical induction."

culations, like inequalities, bottom, and upper limits of infinite series, a subproof by induction, etc. Weierstrass knows the difference between natural and real numbers and is able to simplify expressions with factorials and sums of infinite geometrical series. That this program found the proof completely automatically[33] shows what theorem provers could do if they can make use of various sorts of concepts and calculations. Moreover, the proof is understandable and relatively short. It looks just like a human proof.[34]

5 Conclusion

> Mathematical proof practices—as institutionalized in the profession of mathematics—is ultimately a matter of discovering new results; this is regardless of whether we understand 'why' they are true or not: Mathematics remains theorem-driven. If mathematicians find they can discover more of these results by means of computers than they can by means of traditional proof, they will desert traditional proof for that reason alone.

In this quote, Jody Azzouni claims that understanding has no influence on mathematical proof practices (Azzouni [2005]: 44). I do not agree: Computer provers *have* shown their success by proving some important conjectures. However, it seems that most mathematicians do not use these programs in their daily practice. If Azzouni would be right, a lot more mathematicians would be using computer provers. Why don't they? In this paper I have argued that one of the obstacles lies in the mathematical concepts used in computer proofs. My conclusion is that, contrary to what Azzouni claims, understanding *has* an influence on

---

The inductive proof requires some not-quite-straightforward algebraic manipulations to make use of the induction hypothesis; the program also finds these steps automatically."

[33] The only caveat is that Beeson had to supply an existential variable and a denumerator $q!$, which is a 'hint' for the program to multiply the inequality by $q!$ and then to simplify the expression.

[34] Beeson [1998], "The intention is, to produce a proof that can be read and checked for correctness by a human mathematician; the standard to be met is 'peer review,' just as for journal publication."

mathematical proof practices. The philosopher of mathematics David Corfield pinpoints the same issue:

> What mathematicians are largely looking for from each other's proofs are new concepts, techniques and interpretations. Computer proofs certainly give information concerning the truth of a result, but very little beyond this.
>
> (Corfield [2003]: 56)

The idea that mathematicians are looking for ideas, rather than results, in each other's proofs is also present in an article of computer scientists Manfred Kerber and Martin Pollet:

> One of the reasons why the Principia are so rarely read is that the main ideas of the proofs are no longer visible in very long and very detailed proofs.
>
> (Kerber and Pollet [2002]: 3)

Kerber and Pollet point here to a relation between computer proofs and Russell and Whitehead's *Principia Mathematica*: In both types of proofs, the concepts used are not adapted for understanding. Both computer proofs and the *Principia Mathematica* are mostly used to attain truth, not understanding. If you have to read through 600 to 700 pages before the authors can prove $1 + 1 = 2$, then this doesn't lead to understanding. If you have to read through a telephone book of intricate algebraic manipulations, then this doesn't lead to understanding too. The consequence is that mathematicians have trouble understanding both the *Principia Mathematica* and computer proofs.[35]

Given the observation that many human proofs contain a mix of concepts from different domains, which plays a significant role in giving insight, while many computer proofs nowadays use a limited set of concepts, it is difficult to see how computer proofs can give much insight. From this negative result, we can learn something. Giving computer provers the ability to manipulate a richer set of concepts will have two important consequences: Human mathematicians will understand

---

[35] Of course the Principia is not intended as a textbook to explain arithmetic, but to provide logical foundations for arithmetic. If that takes hundreds of pages, so be it. But with computer proofs, the stakes are different: We don't seek foundations for a theory, but a verification (and preferably also an explanation) of a theorem.

the proofs better, and computers could prove more interesting results. Maybe computer provers could then become part of mainstream mathematical practice.

Acknowledgments

I would like to thank my advisor Leon Horsten, who was invaluable during my Master's thesis about computer proofs and mathematical practice. I also want to thank the organizers of the *Perspectives on Mathematical Practices 2007* conference in Brussels, where I had the opportunity to present the ideas of this paper. Also thanks to various participants of the conference with whom I discussed these ideas, especially to Dirk Schlimm and Mikkel Willum Johansen. Last but not least, I'd like to thank Dirk Schlimm, Herman Roelants, and two anonymous referees for their helpful comments on a draft of this paper.

Bibliography

Aigner, M. and G. Ziegler [2004]. *Proofs from THE BOOK.* 3rd ed. Springer.

Appel, K. and W. Haken [1977a]. "Every planar map is four-colorable: Part I. Discharging." *Illinois Journal of Mathematics* **21**. 429–90.

— [1977b]. "The solution of the four-color map problem." *Scientific American* **237** (4). 108–21.

— [1978]. "The four-color problem." *Mathematics today: twelve informal essays.* Ed. by L.A. Steen. New York: Springer-Verlag. 153–80.

Appel, K.; W. Haken; and J. Koch [1977]. "Every planar map is four-colorable: Part II. Reducibility." *Illinois Journal of Mathematics* **21**. 491–567.

Avigad, J. [2006]. "Mathematical method and proof." *Synthese* **153** (1). 105–59.

Azzouni, J. [2005]. "Is there still a sense in which mathematics can have foundations?" *Essays on the Foundations of Mathematics and Logic.* Ed. by G. Sica. Vol. 1. Advanced Studies in Mathematics and Logic. Monza: Polimetrica. 9–47.

Beeson, M. [1998]. "Automatic derivation of epsilon-delta proofs of continuity." *Lecture Notes in Artificial Intelligence* **1476**. 67–83.
— [2001]. "Automatic derivation of the irrationality of e." *Journal of Symbolic Computation* **32** (4). 333–49.
— [2003]. "The mechanization of mathematics." *Alan Turing: Life and Legacy of a Great Thinker*. Ed. by C. Teuscher. Springer-Verlag. 77–134.
Belinfante, J.G. [1999]. "On computer-assisted proofs in ordinal number theory." *Journal of Automated Reasoning* **22**. 341–78.
Bundy, A. [1988]. "The use of explicit plans to guide inductive proofs." *Lecture Notes in Computer Science* **203**. 111–20.
Burris, S. [1996]. "An anthropomorphized version of McCune's machine proof that Robbins algebras are Boolean algebras." Unpublished manuscript.
Cerutti, E. and P.J. Davis [1969]. "Formac meets Pappus: some observations on elementary analytic geometry by computer." *The American Mathematical Monthly* **76** (8). 895–905.
Colton, S.; A. Bundy; and T. Walsh [1999]. "Automatic concept formation in pure mathematics." *Proceedings of the Sixteenth International Joint Conference on Artificial Intelligence*. 786–93.
— [2000]. "On the notion of interestingness in automated mathematical discovery." *International Journal of Human Computer Studies* **53** (3). 351–75.
Corfield, D. [2003]. *Towards a philosophy of real mathematics*. Cambridge University Press.
Dahn, B. [1997]. *An explanation of EQP/Otter's proof of Winker's second condition for Robbins algebras*. Tech. rep. Unpublished manuscript.
Eagle, A. [2008]. "Mathematics and conceptual analysis." *Synthese* **161** (1). 67–88.
Fitelson, B. [1998]. "Using Mathematica to understand the computer proof of the Robbins conjecture." *Mathematica in Eduction and Research* **7** (1). 17–26.
Fraassen, B. van [1980]. *The Scientific Image*. Oxford: Clarendon Press.

Gonthier, G. [2004]. *A computer-checked proof of the four colour theorem*. Tech. rep. Microsoft Research. URL: http://research.microsoft.com/~gonthier/4colproof.pdf.

Hales, T.C. [2005]. "A proof of the Kepler conjecture." *Annals of Mathematics* **162** (3). 1065–185.

Halmos, P.R. [1990]. "Has progress in mathematics slowed down?" *The American Mathematical Monthly* **97** (7). 561–88.

Huang, T. [2002]. "Automated deduction in ring theory." MA thesis. Department of Computer Science, San Jose State University.

Huntington, E.V. [1933a]. "Boolean algebra. A correction." *Transactions of the American Mathematical Society* **35** (2). 557–8.

— [1933b]. "New sets of independent postulates for the algebra of logic, with special reference to Whitehead and Russell's Principia Mathematica." *Transactions of the American Mathematical Society* **35** (1). 274–304.

Kauffman, L.H. [2001]. "The Robbins problem: computer proofs and human proofs." *Kybernetes* **30** (5/6). 726–51.

Kempe, A.B. [1879]. "On the geographical problem of the four colours." *American Journal of Mathematics* **2** (3). 193–200.

Kerber, M. and M. Pollet [2002]. *On the design of mathematical concepts*. Tech. rep. School of Computer Science, The University of Birmingham. URL: ftp://ftp.cs.bham.ac.uk/pub/authors/M.Kerber/TR/CSRP-02-06.pdf.

Lam, C.W.H.; L. Thiel; and S. Swiercz [1989]. "The non-existence of finite projective planes of order 10." *Canadian Journal of Mathematics* **XLI**. 1117–23.

Manin, Y.I. [1981]. "A digression on proof." *The Two-Year College Mathematics Journal* **12** (2). 104–7.

Martin, U. [1999]. "Computers, reasoning and mathematical practice." *Computational Logic* **56** (3). 346–8.

McCune, W. [1997]. "Solution of the Robbins problem." *Journal of Automated Reasoning* **19** (3). 263–76.

Mumford, D. [1967]. "On the equations defining abelian varieties. III." *Inventiones Mathematicae* **3** (3). 215–44.

Pastre, D. [1999]. *Can and must a machine prove theorems as humans do?* Tech. rep. Université Paris 5. URL: http://www.math-info.univ-paris5.fr/~pastre/CanAndMust.pdf.
Peterson, I. [1997]. "Computers and proof: applying automated reasoning to prove mathematical theorems." *Science News* **151** (12). 178.
Petkovšek, M.; H. Wilf; and D. Zeilberger [1996]. $A = B$. A K Peters.
Rav, Y. [1999]. "Why do we prove theorems?" *Philosophia Mathematica* **7** (1). 5–41.
Resnik, M.D. and D. Kushner [1987]. "Explanation, independence and realism in mathematics." *The British Journal for the Philosophy of Science* **38** (2). 141–58.
Robertson, N. et al. [1997]. "The four-colour theorem." *Journal of Combinatorial Theory, Series B* **70** (1). 2–44.
Robinson, J.A. [1965]. "A machine-oriented logic based on the resolution principle." *Journal of the Association for Computing Machinery* **12**. 23–41.
Sandborg, D. [1998]. "Mathematical explanation and the theory of why-questions." *The British Journal for the Philosophy of Science* **49** (4). 603–24.
Sica, G., ed. [2005]. *Essays on the Foundations of Mathematics and Logic.* Vol. 1. Advanced Studies in Mathematics and Logic. Monza: Polimetrica.
Steen, L.A., ed. [1978]. *Mathematics today: twelve informal essays.* New York: Springer-Verlag.
Teuscher, C., ed. [2003]. *Alan Turing: Life and Legacy of a Great Thinker.* Springer-Verlag.
Thurston, W.P. [1994]. "On proof and progress in mathematics." *Bulletin of the American Mathematical Society* **30** (2). 161–77.
Wos, L. and G.W. Pieper [2003]. *Automated reasoning and the discovery of missing and elegant proofs.* Rinton Press.

ESSAY 5

For a Philosophy of Mathematical Practice

Gianluigi Oliveri

It is a fact that of the logicist, the intuitionist, and Hilbert's programmes, the classical programmes set up to establish mathematics on safe foundations, some have failed, whereas others have long since lost their propulsive force.

Two of the consequences of the crisis of the classical programmes in the foundations of mathematics were: a widespread scepticism towards the possibility of establishing the certainty of mathematical methods, and a new attention paid to mathematical practice.

The main aim of this article is that of individuating some directions of research along which to develop a tenable philosophy of mathematical practice.

1 The Analytic Approach to the Philosophy of Mathematics

The analytic approach to the philosophy of mathematics is called so, because it is most common among analytical philosophers. It is characterized, among other things, by the attention paid to traditional questions concerning the existence and nature of mathematical objects and structures. A prime example of the analytic approach to the philosophy of mathematics at work is, of course, the Dummettian realism/anti-realism dispute.

As is well known, for Dummett, the traditional debate between realists and anti-realists concerning, for instance, number theory can be translated into another type of debate which is about whether it makes sense to say that the number theoretical statements for which we do not have a procedure of decision—these are the elements of the 'disputed class'—are true or false.

According to Dummett, if you want to argue against realism (or anti-realism) concerning number theory, you can just as well argue against the adequacy of a realist (or anti-realist) conception of truth for the number theoretical statements belonging to the disputed class and, therefore, also about whether certain 'forms of deductive arguments ... are to be accepted as valid' (Dummett [1991]: 16) in doing number theory.

If we reflect on what has just been said, we will realize that, in the Dummettian realism/anti-realism dispute, the concepts of meaning, truth, and of mathematical statement loom large. And, of course, these are concepts which are philosophically very important within the analytic approach to the philosophy of mathematics as a consequence of one of the central tenets of analytical philosophy: the linguistic turn.

But, if language and its philosophy prove to be essential to the analytic approach, it must be said that, within this way of doing philosophy of mathematics, the history of mathematics has hardly any rôle to play.

Indeed, traditionally, above all within analytical circles, the history and the philosophy of mathematics interacted only occasionally, and almost accidentally, on specific issues. These were interactions which did not affect the way the two subjects were developed.

Such a state of affairs, far from being the consequence of carelessness or neglect, was explicitly theorized by Frege, the founding father of analytical philosophy, who famously asked:

> Do the concepts, as we approach their supposed sources, reveal themselves in peculiar purity? (Frege [1989]: vii)

And then answered:

> Not at all; we see everything as through a fog, blurred and undifferentiated. It is as though everyone who wished to know about America were to try to put himself back in the position of Columbus, at the time when he caught the first dubious glimpse of his supposed India. Of course, a comparison like this proves nothing; but it should, I hope, make my point clear. It may well be that in many cases the history of earlier discoveries is a useful study, as a preparation for further researches; but it should not set up to usurp their place. (Frege [1989]: vii–viii)

This negative attitude towards the contribution that the history of mathematics can give to the philosophy of mathematics is not to be found only in Frege, and among analytical philosophers, but is something that colours also the thought of Brouwer and Hilbert.

Now, it seems to me that, in contrast with what is usually said about this, such an attitude is not the consequence of the fact that these authors and their programmes are concerned with the foundations of mathematics. Such an attitude has rather to do, on the one hand, with Brouwer's and Hilbert's philosophies of mathematics, for which mathematics is not a science of matters of fact (in Hume's sense), and, on the other, with a too narrow view these thinkers have of the history of mathematics, which is by them seen as a mere temporal ordering of past events bound to develop either into a purely descriptive history of ideas or into an at times entertaining, but philosophically unhelpful, production of biographies.

It must be emphasized that such a way of thinking about the relationship existing between the history and the philosophy of mathematics has not been a strict monopoly of philosophers, but has also been shared by many historians of mathematics as witnessed, even recently, by interesting contributions to the subject such as Moore [1982] and Avellone, Brigaglia, and Zappulla [2002].

2 Back to Mathematical Practice: Wittgenstein

It is interesting to notice that the interpretation of the imperative 'Back to mathematical practice' offered within the analytic approach originates in Wittgenstein's later philosophy.

This becomes clear, if we consider the general strategy used by Wittgenstein in the *Philosophical Investigations* to show that a number of celebrated philosophical questions are nothing but pseudo-problems. Such a strategy consists in challenging the hidden and unwarranted assumption that the meaning of some relevant words is given in terms of their reference, through an appeal to the practice represented by the use of those very words made by the competent speakers.

Were this strategy successful, it would dissolve the philosophical problem of determining whether the reference of those words is empty or not, and, *a fortiori*, the connected philosophical question concerning what kind of entities the referents of those words are.

For example, according to Wittgenstein, the mind/body problem rests on the hidden and unwarranted assumption that certain words like 'pain,' which belong to what has become known as the 'language of sensations,' are denoting. The mind/body problem is in fact, for him, a direct consequence of this assumption, and consists in asking what the nature of the denotation of words like 'pain' is, that is, it consists in asking whether the nature of the denotation of words like 'pain' is *res cogitans* or *res extensa*.

For Wittgenstein, we can dissolve this classical philosophical problem, if, studying how a child learns to use expressions like 'pain,' we are able to show that these expressions are not learned as names for objects, but as linguistic substitutes for natural expressions of pain (like crying, etc.).

With regard to the philosophy of mathematics, we have that the bulk of Wittgenstein's following-a-rule argument appears to be directed precisely against the idea that the meaning of a mathematical rule $\mathfrak{R}$ is given in terms of its reference; and, consequently, also against the idea that understanding the meaning of a mathematical rule $\mathfrak{R}$ consists in grasping the way in which the reference of $\mathfrak{R}$ is presented to us.

An example of the view criticized by Wittgenstein at work is that of considering both the formulae/rules (1) and (2) below, in which n is a natural number greater than 0, as referring to the set of the odd positive integers, which is presented to us in two different ways:

$$2n - 1; \tag{1}$$

$$n \equiv 1 \pmod{2}. \tag{2}$$

According to Wittgenstein, instead:

> [T]here is a way of grasping a rule which is *not* an *interpretation*, but which is exhibited in what we call 'obeying the rule' and 'going against it' in actual cases.
>
> (Wittgenstein [1983]: §I.201)

To this he adds that:

> '[O]beying a rule' is a practice. And to *think* one is obeying a rule is not to obey a rule. Hence it is not possible to obey a rule 'privately': otherwise thinking one was obeying a rule would be the same thing as obeying it.
>
> (Wittgenstein [1983]: §I.202)

For Wittgenstein, grasping a mathematical rule is matter of becoming master of the technique which has been conventionally associated with that rule by the mathematical community. And the mastery of a technique is acquired, according to Wittgenstein, neither through the contemplation of some abstract entity nor as a consequence of the occurrence of some mental state, but by being trained to carry out a certain procedure.

At this point, it is easy to see how, within the context offered by Wittgenstein's philosophy of mathematics, mathematical practice becomes the focus of attention. For, in Wittgenstein's view, mathematical practice is the only efficacious antidote against the bewitching of reason operated by the philosophers' subtle violations of the deep grammar regulating mathematical language-games. Indeed, for Wittgenstein, some violations of the deep grammar of mathematical language-games—violations like those that make us take words like 'one,' 'two,' 'triangle,' 'addition function,' etc. to be names for objects—give origin to the classical philosophical problem concerning the existence and nature of mathematical objects.

It is important to notice here that what Wittgenstein means by 'mathematical practice,' for instance within arithmetic, is not the activity in which mathematicians engage when they search for new arithmetical results or for proofs of arithmetical statements, but more simply the set of techniques we are trained to perform when we learn to add, multiply, etc. In other words, that which Wittgenstein means by 'mathematical practice' is what we, more suggestively, might call 'folk-mathematics.'

But, now, as part of an attempt to assess the merits of Wittgenstein's appeal to mathematical practice, and the influence of this appeal on later developments of the discussion concerning the rôle of mathematical practice within the analytic approach to the philosophy of mathematics, we should consider the following points.

First, although in the following-a-rule argument there is not even a distant hint pointing at the history of mathematics, for Wittgenstein's 'dissolution strategy' to go through, he needs to show that the techniques we are trained to perform when 'we learn to ϕ,' where ϕ is a mathematical rule, are the outcome of convention. But how could he/anybody do this without engaging in some kind of historic-sociological investigation of mathematical practice?

Secondly, if arithmetic—as it seems likely—is not reducible to adding, multiplying, etc. natural numbers, but is, rather, a theory that studies, among many other things, the properties of addition, multiplication, etc. of natural numbers, it follows that considerations relating to how we learn to add, multiply, etc. natural numbers should not prove to be very helpful in a discussion of the natural and legitimate metatheoretical questions concerning, for example, (i) whether Peano arithmetic (PA) produces information about a mathematical structure, and what kind of entity this is; (ii) whether PA is a scientific theory or not; (iii) how the information produced by PA compares with that produced by the empirical sciences; (iv) what is at the root of the objectivity of the statements belonging to PA, etc.

Thirdly, Wittgenstein's position on mathematical practice as a public and objective criterion for the elimination of philosophical pseudo-problems about mathematics inaugurates a season in the philosophy of mathematics during which mathematical activity/practice comes to be considered as something that can and, indeed, must be accounted for independently of philosophical speculation. This way of thinking the relation between philosophy and mathematical practice has led, in recent years, to P. Maddy's mathematical naturalism.

3 Back to Mathematical Practice: Maddy

According to Penelope Maddy, for a mathematical naturalist:

> [M]athematical methodology is properly assessed and evaluated, defended, or criticized, on mathematical, not philosophical (or any other extra-mathematical) grounds.
>
> (Maddy [1998]: 164)

To understand this quotation, we must first clarify what Maddy means by 'mathematical methodology.' For Maddy, mathematical methodology is a set of prescriptions concerning how mathematical practice should be pursued. Well known examples of some such prescriptions are those about whether we should accept the theory of complex numbers as part of mathematics, whether mathematical analysis should be based on infinitesimals, whether we should accept the Axiom of Choice in set theory, etc. Maddy's claim is, therefore, that in modern mathematics the debate about issues such as those just mentioned is not influenced by philosophical considerations.

Maddy's position on the way mathematical methodology should be assessed is very important from a philosophical point of view. For, if Maddy is right then, in contrast with what many thought about this issue, traditional questions concerning the existence and nature of numbers, sets, etc. have no bearing on considerations which have to do with mathematical methodology.

With regard to this issue, the case of Gödel is particularly important. As is well known, even after Cohen had proved the independence of the continuum hypothesis (CH) from the axioms of Zermelo Fraenkel Choice (ZFC), Gödel, motivated by mathematical Platonism, still thought that mathematicians should have searched for new axioms to be added to the axiomatic basis of ZFC which, by bringing to the fore new relevant aspects of set theoretical reality, would have helped to decide CH.

Maddy's position on mathematical methodology is reiterated in the following quotation where she also draws an interesting and telling distinction between what she means by 'mathematical naturalism' and naturalism about science:

> [N]aturalism about mathematics—Maddy says—differs markedly from naturalism about science in its treatment of philosophical considerations. Consider, for example, the typically metaphysical claim that mathematical objects exist objectively and non-spatiotemporally. The mathematical naturalist holds these issues to be external to mathematics proper and thus irrelevant to methodological decision-making, but the analogous claim about physical objects—that they exist objectively and spatiotemporally—is part and parcel of scientific thinking. This is not true of the corresponding mathematical questions, which is why the mathematical naturalist undertakes to eliminate them from methodological arguments.
>
> For these reasons, the mathematical naturalist pursuing questions of methodology ignores traditional philosophical questions such as 'are mathematical things objective or subjective?', 'is their existence dependent on our theories or definitions?', 'are mathematical objects incomplete?', 'are they more like fictional objects or physical objects?', 'are the axioms true in the real world of sets?', and so on.
>
> (Maddy [1998]: 172–3)

One feature of interest for us here is that, in contrast with the tradition inaugurated by Frege, the history of mathematics is given, through the work of Maddy, an important rôle to play within the analytic approach to the philosophy of mathematics.

Indeed, if we want to engage in a serious assessment of mathematical methodology, in the sense of Maddy, we cannot do this without an appeal to the history of mathematics. That this is the case is shown, for instance, by the consideration that it is indispensable, to an assessment of the criteria according to which the mathematical community eventually decided to eliminate the infinitesimals from mathematical analysis in favour of the theory of limits, to proceed to a preliminary historical reconstruction of the controversy that raged (for a long time) over the use of infinitesimals in analysis.

However, going back to the topic of mathematical practice, it is important to notice that the difference between Maddy and Wittgenstein on this issue is very clear. The examples of methodological prescriptions listed above show that what Maddy means by the 'practice of modern mathematics' is not—in contrast with what Wittgenstein believes to be the case—what we are trained to do, for example, in arithmetic when we learn to add, multiply, etc. natural numbers, but is, rather, what the number theorist does when he contributes to his subject.

The difference between Wittgenstein's folk-mathematics and Penelope Maddy's characterization of mathematical practice is very important, because, as we have already seen, both these views perform an important philosophical rôle, and, moreover, result in very different pictures of mathematics.

If, in the light of what I have already argued, the first point above—on the importance of the difference existing between Wittgenstein's and Maddy's characterization of mathematical practice—needs no further elaboration, I have to dwell a little longer on the second.

Wittgenstein's view of mathematical practice in his later work suggests a picture of mathematics as some kind of social game belonging to the folklore of a certain society.

In the *Philosophical Investigations* Wittgenstein has given up on the hardness of the logical must as what objectively dictates and explains the correctness of a certain way of following-a-mathematical-rule. On the contrary, a particular practice solidifies into a logical must, for him, because it comes to occupy—as a consequence of convention—a particularly central position within what we could describe, borrowing the expression from *On Certainty*, as our *system of knowledge.*

In the case of Maddy, we receive, instead, the impression that she does not intend to undermine the traditional, convention-independent, rôle of what I have repeatedly called the 'logical must.' She simply brackets away the problem of clarifying the origin and nature of such a concept, because she thinks that this is irrelevant to an assessment of mathematical methodology. With Maddy, mathematical methodology becomes one of mathematics' internal affairs.

At this point one might notice some kind of inconsistency in Maddy's thought between the act of bracketing away the problem concerning the origin and nature of the logical must, and the rôle that the history of mathematics plays in questions relating to the assessment of mathematical methodology. That the one above is not a problem for Maddy is shown by the observation that, according to her, clarifying the origin and nature of the logical must is a philosophical, and not a mathematical, problem.

However, in spite of all the differences, there also is an important analogy between Wittgenstein's and Maddy's conceptions of mathematical practice. To see this consider that, according to Wittgenstein, mathematical practice—what I have called 'folk-mathematics'—is innocent of philosophical contamination; and that, for Maddy, not only mathematical practice, but also the metatheoretical methodological debate about mathematical practice, must be 'properly assessed and evaluated, defended, or criticized, on mathematical, not philosophical (or any other extra-mathematical) grounds.'

If, in concluding this section, we attempt to give a critical evaluation of Wittgenstein's and Maddy's views of mathematical practice, we find that they are both unsatisfactory, though on different accounts. And, since I have already justified my opinion concerning Wittgenstein's position, in what follows I am going to concentrate on explaining my view of Maddy's.

It seems to me that the main defect of Maddy's idea of mathematical practice coincides with the distinguishing mark of what she calls 'mathematical naturalism.' What I mean by this is that Maddy's opinion that 'mathematical methodology is properly assessed and evaluated, defended, or criticized, on mathematical, not philosophical (or any other extra-mathematical) grounds' is evidence of a very narrow and essentially inadequate view of mathematical theories and their history, as is shown by the large number of important counterexamples present within contemporary mathematics.

One of the best such counterexamples is represented by the work of that strange, many-headed French mathematician known under the name of Nicolas Bourbaki.

In the huge production of Nicolas Bourbaki we come across an eminently philosophical idea about the nature of mathematics, according to which mathematics is a science of structures. This is an idea with vast and profound consequences on mathematical methodology and practice, because it inspires and guides Bourbaki's truly heroic attempt to rewrite and systematize the whole of mathematics in its light.

Another substantial counterexample to Maddy's ideas on mathematical methodology and practice is given by the challenge recently posed by category theory to set theory as the proper foundation for mathematics. This is another important case in which philosophical ideas on the existence and primitive nature of sets vs. algebra play a central rôle in motivating and directing not only a philosophical controversy, but a large amount of mathematical research as well.

With regard to this last point, we should notice that it does not carry much weight replying to the above counterexample saying 'Yes, perhaps that is what has happened, but they should have not done so.' For, there is no denying that considerations concerning the existence and primitive nature of sets vs. algebra (i) are part of the mathematician's view (philosophy) of his subject;[1] and that (ii) they are also inseparable from his 'technical' choice.

What I mean by (ii) above is that, independently of the degree of philosophical awareness, and of the explicit motivations, of the mathematician, were he to show that set theory is more fundamental than category theory (or *vice versa*) this would have profound and objective philosophical consequences.

---

[1] It is a genuine philosophical question to ask: What is the basic 'furniture' of mathematics?

To see this consider the obvious application of Ockham's razor to categories (or sets) resulting from his work. This is an application which would have a clear bearing on genuine philosophical questions like: What is mathematics really about?

But, of course, if, as I claimed, there are branches of mathematics in which the mathematician's technical choice about methodology, i.e., whether to go for set theory or category theory as a foundation for mathematics, is inseparable from philosophical considerations, it follows that these branches of mathematics are bound to be good counterexamples for Maddy's mathematical naturalism.

For, in cases such as these, being mathematical methodology strongly influenced by the mathematician's view of his subject, it cannot be properly assessed and evaluated, defended, or criticized, (only) on mathematical, not philosophical grounds.

4 The Historical Approach to the Philosophy of Mathematics

What I have described at the beginning of the paper (in §2) as the traditional separation between the history and the philosophy of mathematics began to be challenged by I. Lakatos in *Proofs and Refutations*.

In much of Lakatos's work in the philosophy of mathematics, the history of mathematics is no longer seen as a mere descriptive enterprise, but becomes the battle ground for rational reconstructions guided by explicit philosophical paradigms such as the Popperian one of conjectures and refutations. These are philosophical paradigms which provide explicit schemata of interpretation of an, otherwise, unreconstructed mass of data.

The traditional Lakatosian approach to the philosophy of mathematics originates from the observation that abstract speculations concerning what mathematical theories are (or should be), that is, speculations which bear no relevance to, or are irreconcilable with, mathematical practice, are, to paraphrase Kant, empty; and that considerations concerning the nature of mathematics which consist simply of an unreconstructed mass of temporally ordered data are blind.

Secondly, in contrast with the analytic standpoint, the traditional

Lakatosian approach to the philosophy of mathematics is not interested in problems concerning the existence of mathematical objects, etc., but tries, instead, to address questions of demarcation between mathematics and the empirical sciences focussing neither on the nature of mathematical statements—whether they are analytic or synthetic, *a priori* or *a posteriori*—nor on the structure of mathematical theories, but on the study of how mathematical knowledge grows.

Lastly, for Lakatos, the term 'mathematical practice' refers to the product of the mathematician's activity when he contributes to his subject, and not, therefore, to anything like Wittgenstein's folk-mathematics. On the other hand, according to him, the contributions given by a mathematician to his subject are not simply those involved in proving theorems, but extend also to the heuristic side of the mathematician's work, a side which is essential to what Lakatos calls 'informal mathematics.'

This latter aspect of Lakatos's view of mathematical practice is very important. For, considering the so-called 'context of mathematical discovery'—besides marking a clear point of departure from the traditional justification-oriented representation of the mathematician's activity—gives a more complete picture of mathematical practice, and reveals that the history of mathematics is indispensable to its study.

5 On the Internal and External History of a Mathematical Theory

But, now, claiming that the history of mathematics must be at the heart of an acceptable way of debating philosophical questions concerning mathematics, is not to say that the history of mathematics provides a philosophically neutral tribunal before which philosophical problems must be brought.

In fact, since the history of mathematics does not consist of a mere chronology of events, but of a rational reconstruction of the way mathematical knowledge grows; and since this rational reconstruction is attained by means of selective attention paid to mathematical events which are distinguished into causes, effects, etc. according to a certain theory, it follows that the history of mathematics is, to a large extent, the outcome of the interpretation, and of the evaluation of certain events obtained through the use of concepts and ideas which originate within the philosophy of mathematics.

Therefore, I agree with Lakatos for whom:

> [T]he historiography of [mathematics] should learn from the philosophy of [mathematics] and *vice versa*.
> (Lakatos [1971]: 102)

The assertion above has the ring of truth about it, and shows in very clear terms another strong disanalogy between Lakatos's philosophy of mathematics and Maddy's naturalism.

However, if it is straightforward to see why the philosophy of mathematics should learn from the historiography of mathematics, it is, nevertheless, legitimate to ask what exactly does the philosophy of mathematics contribute to the historiography of mathematics.

A very compact answer to this problem is that the

> [P]hilosophy of [mathematics] provides normative methodologies in terms of which the historian reconstructs 'internal history' and thereby provides a rational explanation of the growth of objective knowledge[.] (Lakatos [1971]: 102)

To make it possible to understand the above answer, I must explain what the internal and external histories of a mathematical theory are. The internal history of a mathematical theory is the account of how the mathematical theory has evolved in relation to the set of its problems.

A very important feature of the internal history of a mathematical theory is that it is normative, because the particular shape that this takes depends on the methodology you adopt to account for historical events.

To see this consider what happened in analysis when the infinitesimals were eliminated. If your methodology allows (normative aspect) theory-refutations, then you might explain the elimination of infinitesimals from analysis as an instance of the refutation of a mathematical theory. If, on the other hand, your methodology does not allow theory-refutations then you might say that the elimination of the theory of infinitesimals from analysis is an example of the elimination of misguided philosophical ideas which had been introduced within analysis when mathematicians had not yet hit against the genuinely mathematical concepts which give a correct account of differentiation, integration, continuity, etc.

It is this normative function performed by methodology in shaping up the internal history of a mathematical theory what prompts me to call such a methodology 'logic of mathematical discovery,' with the proviso that the use of the term 'discovery' in the expression above is purely metaphorical, and should not, therefore, suggest any realist bias in the understanding of the nature of mathematical activity.

Concerning the external history of a mathematical theory, I must point out that this is the account of the psycho-socio-cultural conditions relevant to the development of a given mathematical theory.

Interesting examples of external history at work are Mancosu's [1996] discussion of the way Leibniz's and Newton's ideas on the new calculus were received by the French Academy of Sciences and Gray's [2006] recent argument to the effect that modern mathematics is a particular case of what he calls 'modernism.'

With regard to the rôle of external history, I must say with Lakatos that

> [A]ny rational reconstruction of history needs to be supplemented by an empirical (socio-psychological) 'external history.' (Lakatos [1971]: 102)

The reason for this is that only the external history of a mathematical theory is capable of explaining why that theory was developed in a particular country at a particular time; the speed with which certain ideas were accepted; etc. by making explicit the psycho-socio-cultural factors which, together with the demands made by the economy on mathematics in terms of technological applications, play the very important rôle of catalysts.

Now that I have clarified the concepts of internal and external history of a mathematical theory, it is important to address the following question. If the philosophy of mathematics contributes normative methodologies to the historiography of mathematics, normative methodologies which are used by the historian to reconstruct the internal history of a mathematical theory, what happens when two different normative methodologies come into conflict with one another?

The answer to this question is that, as in the case concerning the normative methodologies (inductivism, falsificationism, etc.) adopted within the empirical sciences to write their internal histories, there has been a rational debate which has led to the rejection of some of these normative methodologies and to the acceptance of others; in the same way, in the case offered by mathematical theories, it is possible to come to a rational decision about which among two competing normative methodologies is the best by examining the consequences of the adoption of these methodologies on normatively interpreted history.

Lakatos's ideas, although very stimulating, present us with a number of serious problems which show that Lakatos's philosophy of mathematics, as originally formulated, is untenable. Here there is room for discussing only one such a problem.

According to Lakatos, the distinction between formal and informal mathematics (see §5) is a distinction in which informal mathematics performs the rôle of some kind of reality, which is the object of description of formal mathematics. That this is not a satisfactory representation of the relationship existing between formal and informal mathematics is shown by the consideration that whereas reality is incorrigible by definition, informal mathematics is not. Informal mathematics is often corrected by the findings and procedures of formal mathematics.

Indeed, for example, informal, or naïve, set theory had to be modified by axiomatic versions of set theory as a consequence, among other things, of the contradictions; and the introduction of the ϵ, δ-proofs in analysis was the consequence of Cauchy's and Weierstrass's concern and investigations about rigour (see Grabiner [1981]).

This is a very important point against the original Lakatosian view of mathematics. For, if informal mathematics is modified by formal mathematics, then informal mathematics cannot be thought as the source of *heuristic falsifiers*[2] for formal theories, and, consequently, the *conjec-*

---

[2]See Lakatos [1983a]: 36,

> [I]f we insist that a formal theory should be the formalization of some informal theory, then a formal theory may be said to be 'refuted' if one

tures and refutations paradigm that is at the basis of the Lakatosian logic of mathematical discovery is undermined.

6 A Philosophy of Mathematical Practice

Having examined both the analytic and historical approaches to mathematical practice, time has now come to produce some indication concerning the research directions following which it might be possible to develop a tenable philosophy of mathematical practice.

But before I do so, I should, perhaps, spell out some of the principles around which an acceptable philosophy of mathematical practice should revolve; and warn the reader that the justifications I am going to offer here for the introduction of such principles are purely 'genetic,' in the sense that they derive directly from my discussion of Wittgenstein's, Maddy's, and Lakatos's standpoints on mathematical practice.

One of the most important principles that a tenable philosophy of mathematical practice should be guided by is something we find in the views of mathematical practice held by the three above mentioned authors. This is the idea that mathematical practice does not stand in need of a justification either from higher—logico-metaphysical—or from lower—psycho-socio-cultural—assumptions, but that it, rather, must become the object of an elucidation.

This principle is, on the one hand, a consequence of a reflection on the failure of the traditional foundational programmes, and of the common view that any reference to psychology, sociology, and culture is ultimately unable to provide a satisfactory account of phenomena which very strongly characterize mathematical practice such as that of the objectivity of mathematical statements; and, on the other, is the outcome of the observation that much conceptual clarification is needed in order to eliminate pseudo-problems, assess mathematical methodology, and reveal the true logic of mathematical discovery.

Moreover, we ought to observe that, although the principle above turns out to be incompatible with any temptation to produce global

---

of its theorems is negated by the corresponding theorem of the informal theory. One could call such an informal theorem a *heuristic falsifier* of the formal theory.

justifications of mathematical practice, this is neither opposed to the axiomatic method as such nor, consequently, to the study of the foundations, that is, of the axiomatic basis, of mathematical theories.

Another principle that an acceptable philosophy of mathematical practice should adopt can be extrapolated from Lakatos. According to this principle, mathematical practice must be considered as a basis for the investigation of its internal and external preconditions.

To see what this means consider that, if the fragment of mathematical practice we are interested in is Cantor–Zermelo set theory (CZST), the principle above would prompt us to investigate, as internal preconditions, which concept of infinity (actual), techniques of proof (diagonalization), concept of set (iterative), part/whole relation, logic, etc. must we adopt to have CZST; and, as external preconditions, this principle would lead us to study questions that are relevant to the context in which CZST appeared and developed, that is, questions like: the use of infinity in traditional pre-Cantorian mathematics, the issue about the relationship between infinity and God in Christian theology at the end of the nineteenth century, the formation, at around the same time, of pressure groups with an influence on academic life, etc.

We should notice that the principle above is justified by the Lakatosian presupposition of the existence of an internal and an external history of mathematical theories, and by the fact that both within mathematical theories and the psycho-socio-cultural context in which mathematical theories develop, it is possible to individuate a nucleus of notions/factors which perform the rôle of preconditions, in the sense that they provide a framework for the others.

According to a further principle that we can abstract from Lakatos, the study of mathematical practice must be carried out by means of an unashamedly philosophically-laden history of mathematics. This is a history of mathematics which should be able to delineate accounts of the internal and external history of mathematical theories which are in harmony with one another.

The introduction of this principle is justified by the consideration that any history of mathematics is philosophically laden, and that different philosophical ideas used to produce rational reconstructions of the development of mathematical theories result in different normative methodologies, which, implemented in the history of mathematics, give different accounts of the growth of mathematical knowledge.

A fourth principle that is not found in any of the authors I have considered so far, but which must be adopted by a philosophy of mathematical practice that aims at challenging the established standpoint in the philosophy of mathematics, is this: A tenable philosophy of mathematical practice should produce some progress on problems belonging to what I have called the 'established standpoint in the philosophy of mathematics.'

Indeed, if with the introduction of a philosophy of mathematical practice no such a progress were forthcoming, then, all being equal, there would be no truly logically compelling reason for rejecting the established standpoint, and accepting in its place a philosophy of mathematical practice.

Although the list of principles offered above is by no means exhaustive, I hope that (i) the justifications produced for their adoption make sufficiently clear why I think that these principles ought to be at the heart of a tenable philosophy of mathematical practice and that (ii) a philosophy of mathematical practice revolving around the principles above operates a synthesis of the analytic and historical approaches to the philosophy of mathematics.

Concerning the latter point, we must notice that the first principle—that mathematical practice does not stand in need of justifications, but only of an elucidation—captures a feature that, as we have already seen, is common to both the analytic (as represented by the views of Wittgenstein and Maddy), and the historical approaches (Lakatos) to the philosophy of mathematics.

The second principle—that mathematical practice must be considered as a basis for the investigation of its internal and external preconditions—absorbs within itself the analytic interest in the so-called 'internal preconditions' and, therefore, in the internal history of a mathematical theory, together with the attention for the external history, and preconditions, of the theory, which are of paramount importance for the historical approach to the philosophy of mathematics.

The third principle—that the study of mathematical practice must be carried out by means of a philosophically-laden history of mathematics—clearly conflicts with the views of Wittgenstein and Maddy. For, according to Wittgenstein, the history of mathematics has no rôle to play within the philosophy of mathematics; and, for Maddy, the history of mathematics useful to the study of mathematical methodology is not philosophically-laden.

However, this is not a problem for the claim that the philosophy of mathematical practice here sketched operates a synthesis of the analytic and historical approaches to the philosophy of mathematics, because, as I have already argued, Wittgenstein's and Maddy's positions on the nature and rôle of the history of mathematics within the philosophy of mathematics are erroneous.

Lastly, the fourth principle—that a tenable philosophy of mathematical practice should produce some progress on problems belonging to the established standpoint in the philosophy of mathematics—speaks of the synthetic, ecumenical aim of a philosophy of mathematical practice which is manifested in the desire to make one's own the problems of other approaches to the philosophy of mathematics.

What I aim to do in this final part of the paper, is producing an outline of a normative methodology obtained implementing the principles above, and showing that its successful application to the history of mathematics could produce, among other things, some progress with regard to the traditional realism/anti-realism debate in the philosophy of mathematics.

Let us immediately say that the normative methodology at issue here is an adaptation for mathematics of the Lakatosian Methodology of Scientific Research Programmes (MSRP).[3] The next question, of course, is: What is a Lakatosian Scientific Research Programme (SRP)?

A scientific research programme is, in Lakatos's view, a finite sequence of theories developing from an original hard core according to

---

[3] As is well known, the attempt to apply the Lakatosian MSRP to the philosophy of mathematics has been defended by some (see Hallett [1979a]; Hallett [1979b]; Oliveri [1997]; Oliveri [2006]; and Oliveri [2007]), and criticized by others (see Koetsier [1991] and Corfield [2003]). However, given (i) the complex nature of the controversy presently raging over this issue, (ii) the aim, and (iii) the already considerable length of the present paper, I suggest that the interested reader should consult on these matters Oliveri [2007], where he can find a discussion of some of the main issues involved (§6), and a successful application to Cantor–Zermelo set theory of a modified version of such a normative methodology (§7).

certain methodological rules. Some of these rules belong to the so-called 'negative heuristic'; whereas the others, not surprisingly, belong to the 'positive heuristic.'[4]

Whereas the negative heuristic specifies the hard core of the research programme,[5] and forbids researchers to refute assumptions belonging to it; the positive heuristic, instead, leads researchers towards the construction of auxiliary hypotheses which are useful to (i) eliminate the anomalies resulting from an application of the research programme made in terms of predictions, etc., and (ii) solve puzzles generated by the introduction of the research programme. Lakatos calls such a set of auxiliary hypotheses the 'protective belt' of the hard core.[6]

According to Lakatos, SRPs can be either *progressive* or *degenerating*:

> [A] series of theories $[T_1, \ldots, T_n$ which expresses an SRP] is *theoretically progressive* if each new theory has some excess empirical content over its predecessor, that is, if it predicts some novel, hitherto unexpected fact . . . a theoretically progressive series of theories is also *empirically progressive*

---

[4]Lakatos [1983b]: 47,

> The programme consists of methodological rules: some tell us what paths of research to avoid (*negative heuristic*), and others what paths to pursue (*positive heuristic*).

[5]Lakatos [1983b]: 49–50,

> Research programmes, besides their negative heuristic, are also characterized by their positive heuristic . . . The negative heuristic specifies the 'hard core' of the programme which is 'irrefutable' by the methodological decision of its proponents; the positive heuristic consists of a partially articulated set of suggestions or hints on how to change, develop the 'refutable variants' of the research-programme, how to modify, sophisticate, the 'refutable' protective belt.

[6]Lakatos [1983b]: 48,

> All scientific research programmes may be characterized by their '*hard core*.' The negative heuristic of the programme forbids us to direct the *modus tollens* at this 'hard core.' Instead, we must use our ingenuity to articulate or even invent 'auxiliary hypotheses', which form a *protective belt* around this core, and we must redirect the *modus tollens* to *these*.

> if some of this excess empirical content is also corroborated, that is, if each new theory leads us to the actual discovery of some *new fact*. Finally, let us call [an SRP] *progressive* if it is both theoretically and empirically progressive, and *degenerating* if it is not. (Lakatos [1983b]: 33–4)

The distinction between *progressive* and *degenerating* SRPs is very important, because it provides a rational criterion of choice between opposing and incompatible SRPs.

Lastly, for Lakatos, the finite sequence of theories $T_1, T_2, \ldots, T_n$, which characterizes a given progressive SRP, is made of falsifiable theories all of which but the last have been falsified.

It is important to notice that, according to Lakatos:

> [A] scientific theory T is *falsified* if and only if another theory T' has been proposed with the following characteristics: (1) T' has excess empirical content over T: that is, it predicts *novel* facts, that is, facts improbable in the light of, or even forbidden, by T; (2) T' explains the previous success of T, that is, all the unrefuted content of T is included (within the limits of observational error) in the content of T'; ... (3) some of the excess content of T' is corroborated[.]
> (Lakatos [1983b]: 32)

Also: (4) T' must be able to provide predictions/explanations of events for the predictions/explanations of which T' has *not* been devised (Zahar's criterion; see Lakatos and Zahar [1983]).

We must notice that, for Lakatos, the *falsifiability* of theories belonging to a given progressive SRP, shows that the criteria of choice adopted by the scientific community *ultimately* rest on a rational evaluation of properties of the theories rather than on a mood or a fashion.[7] The same considerations apply to the choice made by the scientific community between two opposing and mutually incompatible SRPs, a choice based on the *progressive/degenerating* distinction.

---

[7]Lakatos would then resist positions such as Ernest's *Social Constructivism* (see on this Ernest [1997]) and Glas's ([1995]: 225) "'post-positivist' [view of mathematics which emphasizes] cognitive practices drawing on socially shared and socially transmitted goals, methods, standards and criteria."

When we turn to the question of applying some of Lakatos's ideas concerning the MSRP to the philosophy of mathematics, we come across a number of familiar, very serious problems, e.g., 'What is the content of the hard core of a would-be mathematical research programme?' 'How can we formulate a satisfactory concept of fallibility for mathematical theories?' *Et cetera.*

The nature of these problems is such that it requires a radical change of the original Lakatosian framework for this to become applicable to the philosophy of mathematics. For instance, if we want to produce an adaptation for the philosophy of mathematics of the Lakatosian key notion of *falsifiability*, we must realize that, in the case of mathematical theories, 'excess empirical content' of a theory T' with respect to a theory T ought to be interpreted as: results which are provable in T' but not in T;[8] condition (2)—that all the unrefuted content of T is included in T'—can be taken literally; condition (3)—that some of the excess content of T' is corroborated—is not applicable except in the sense that the theory T', in which it is possible to prove results which are not provable in T, has a model; and condition (4)—that T' has not been obtained by means of the introduction into T of *ad hoc* assumptions—like condition (2), can be taken literally.[9]

The above reinterpretation of the *falsifiability* conditions, which renders them applicable to mathematical theories, shows, together with other important changes to be made to the original Lakatosian framework, the existence of substantial differences between mathematical theories and theories belonging to the empirical sciences, and justifies us

---

[8] I am not here considering either experimental or applicable mathematics to explain how to apply to mathematical theories an adequate translation of the excess empirical content relation holding between two scientific theories, because: First, it is the bulk of its theorems what I take the content of a mathematical theory to be; and, secondly, at any one time the terms 'experimental' and 'applicable' refer only to some mathematical theories, and not to others whereas what is required here is a criterion that can be applied to any two mathematical theories belonging to the same research programme.

[9] Notice that conditions (3) and (4) greatly restrict the possibility of producing extensions of mathematical theories. The most glaring case of *ad hocness*—if I am allowed the contamination of Latin with English—being that which obtains when you add to the set Σ of axioms of a mathematical theory T a proposition ϕ, which is not known to be independent of the axioms of T, with the sole purpose of proving ϕ from $\Sigma \cup \{\phi\}$.

in talking about Mathematical Research Programmes (MRPs), rather than about Scientific Research Programmes, when dealing with mathematical theories. It, furthermore, suggests an obvious extension of the *progressive/degenerating* distinction to MRPs, a distinction drawn along the following lines: an MRP is *theoretically progressive* if there are results provable in each new theory belonging to it, which are not provable within its predecessor; an MRP is *factually progressive*, if each new theory has a model. An MRP is *progressive*, if it is both *theoretically* and *factually progressive*, and *degenerating*, if it is not.

Concerning the normative methodology I have just presented, we need to notice that this develops in the respect of the four principles of a philosophy of mathematical practice singled out above.

In fact, the aim of the Methodology of Mathematical Research Programmes (MMRPs) is not that of justifying mathematical practice, but the very different one of elucidating it by means of its rational reconstructions.

Secondly, by means of the hard core/protective belt distinction, the MMRP is particularly suitable for an investigation of the internal preconditions of a mathematical theory, because these may be identified as the elements of the hard core of the MRP.

Thirdly, the rational reconstructions of episodes belonging to mathematical practice, by means of an application of the MMRP, result in the writing of a strongly philosophically-laden history of mathematics.

For, as we have already seen, the MMRP itself is the outcome of an explicit and deliberate philosophical reflection on mathematical practice and its philosophical relevance.

Fourthly, an application of the progressive/degenerating distinction to two rival MRPs, one of which has realist assumptions in its hard core, whereas the other's hard core contains assumptions of an anti-realist kind, might rationally decide a traditional metaphysical issue—an issue very dear to the analytic approach to the philosophy of mathematics—in favour of the metaphysical assumptions belonging to the hard core of the progressive MRP by means of concepts and techniques developed within the historical approach to the philosophy of mathematics.

However, at this point, someone might be tempted to object that, if we consider the Cantor–Zermelo MRP in which CH is assumed to be true, and the Cantor–Zermelo MRP in which ¬CH is assumed to be true, we have here two MRPs which cannot be compared by means of the progressive/degenerating distinction, because CH is independent of the axioms of ZFC.[10]

Now, it seems to me that the objection above is based on the mistaken assumption that Cantor–Zermelo set theory + CH and Cantor–Zermelo set theory + ¬CH are two different MRPs. Indeed, the problem posed by the objection is dissolved once we realize that CH is a component of the protective belt of the Cantor–Zermelo MRP, a component devised to sort out a puzzle arising from the introduction of this MRP: How many different types of infinity are present in $\mathbb{R}$?

Therefore, since the formal systems ZFC + CH and ZFC + ¬CH share the same hard core together with its realist assumptions, they undermine neither the progressive/degenerating character of the Cantor–Zermelo MRP, nor the realist commitment of its hard core.

Furthermore, even if we disregard the possibility, envisaged by Gödel, that the discovery/introduction of a new axiom might decide CH—an example of such an axiom might be $V = L$—we could always say that, since ZFC + CH and ZFC + ¬CH belong to the Cantor–Zermelo MRP, they could be seen, at the very most, as elements of sub-MRPs of the Cantor–Zermelo MRP.

However, the existence within the same MRP of different sub-MRPs would neither be an obstacle to an application of the progressive/degenerating distinction to the sub-MRPs (along the lines indicated in this paper), nor to MRPs. For, we could say that an MRP $\mathcal{K}$ is *progressive*, if there exists at least one sub-MRP of $\mathcal{K}$ that is progressive, and *degenerating* otherwise.

The point above concerning the possibility of applying the progressive/degenerating distinction to two rival MRPs is very important, because the application of the MMRP to mathematical practice, besides producing some progress with regard to an important open question belonging to the analytic school of thought about mathematics, shows that an acceptable philosophy of mathematical practice does not have to have the suicidal approach to mathematical practice advocated by philosophers like Wittgenstein and Maddy. This, as we know, is the

---

[10] A very similar objection was communicated to me by S. Meyer.

approach whereby genuine metaphysical speculations about mathematics either become the outcome of grammatical errors or turn out to be irrelevant to discussions relating to mathematical methodology.

Indeed, an acceptable philosophy of mathematical practice can keep discussing traditional metaphysical issues about mathematics with the hope of producing logically compelling results, if it starts from mathematical practice and, 'working backwards,' succeeds in unveiling its metaphysical assumptions.

Acknowledgments

I wish to thank D. Gillies, S. Meyer, B. Van Kerkhove, and an anonymous referee for having read, and commented upon, a previous version of this paper. I am particularly grateful to B. Van Kerkhove for having invited me to speak at the truly excellent Perspectives on Mathematical Practices conference held in Brussels from the 26th to the 28th of March 2007, and to Jonas De Vuyst for the accurate copy-editing of the article.

Bibliography

Aspray, W. and P. Kitcher, eds. [1988]. *History and Philosophy of Modern Mathematics.* Minnesota: University of Minnesota Press.

Avellone, M.; A. Brigaglia; and C. Zappulla [2002]. "The Foundations of Projective Geometry in Italy from De Paolis to Pieri." *Archive for History of Exact Sciences* **56**. 363–425.

Corfield, D. [2003]. *Towards a Philosophy of Real Mathematics.* Cambridge: Cambridge University Press.

Dales, H.G. and G. Oliveri, eds. [1998]. *Truth in Mathematics.* Oxford: Oxford University Press.

Dummett, M.A.E. [1991]. *The Logical Basis of Metaphysics.* London: Duckworth.

Ernest, P. [1997]. "The Legacy of Lakatos: Reconceptualising the Philosophy of Mathematics." *Philosophia Mathematica* **5** (3). 116–34.

Ferreirós, J. and J.J. Gray, eds. [2006]. *The Architecture of Modern Mathematics: Essays in History and Philosophy.* Oxford: Oxford University Press.

Frege, G. [1989]. *The Foundations of Arithmetic.* Trans. by J.L. Austin. Second Revised Edition. First published in 1884. Oxford: B. Blackwell.

Gillies, D., ed. [1995]. *Revolutions in Mathematics.* Oxford: Clarendon Press.

Glas, E. [1995]. "Kuhn, Lakatos, and the Image of Mathematics." *Philosophia Mathematica* **3** (3). 225–47.

Grabiner, J.V. [1981]. *The origins of Cauchy's rigorous calculus.* Cambridge, MA: MIT Press.

Gray, J.J. [2006]. "Modern mathematics as a cultural phenomenon." *The Architecture of Modern Mathematics: Essays in History and Philosophy.* Ed. by J. Ferreirós and J.J. Gray. Oxford: Oxford University Press. 371–96.

Grosholz, E. and H. Breger, eds. [2000]. *The Growth of Mathematical Knowledge.* Dordrecht: Kluwer Academic Publishers.

Hallett, M. [1979a]. "Towards a Theory of Mathematical Research Programmes (I)." *British Journal for the Philosophy of Science* **30**. 1–25.

— [1979b]. "Towards a Theory of Mathematical Research Programmes (II)." *British Journal for the Philosophy of Science* **30**. 135–59.

Koetsier, T. [1991]. *Lakatos' Philosophy of Mathematics: A Historical Approach.* Amsterdam: North-Holland.

Lakatos, I. [1971]. "History of science and its rational reconstructions." *Philosophical Papers.* Vol. I. Cambridge: Cambridge University Press. 102–38.

— [1983a]. "A Renaissance of empiricism in the recent philosophy of mathematics?" *Philosophical Papers.* Vol. II. Cambridge: Cambridge University Press. 24–42.

— [1983b]. "Falsification and the Methodology of Research Programmes." *Philosophical Papers.* Vol. I. Cambridge: Cambridge University Press. 8–101.

— [1983c]. *Philosophical Papers.* Cambridge: Cambridge University Press.

Lakatos, I. and E. Zahar [1983]. "Why did Copernicus's research programme supersede Ptolemy's?" *Philosophical Papers.* Vol. I. Cambridge: Cambridge University Press. 168–92.

Maddy, P. [1998]. "How to be naturalist about mathematics." *Truth in Mathematics*. Ed. by H.G. Dales and G. Oliveri. Oxford: Oxford University Press. 161–80.
Mancosu, P. [1996]. *Philosophy of Mathematics & Mathematical Practice in the Seventeenth Century*. Oxford: Oxford University Press.
Moore, G.H. [1982]. *Zermelo's Axiom of Choice*. New York, Heidelberg, and Berlin: Springer Verlag.
Oliveri, G. [1997]. "Criticism and Growth of Mathematical Knowledge." *Philosophia Mathematica* **5** (3). 228–49.
— [2006]. "Mathematics as a Quasi-Empirical Science." *Foundations of Science* **11**. 41–79.
— [2007]. *A Realist Philosophy of Mathematics*. London: College Publications.
Tymoczko, T., ed. [1986]. *New Directions in the Philosophy of Mathematics*. Boston: Birkhäuser.
Wittgenstein, L. [1979]. *On Certainty*. Oxford: B. Blackwell.
— [1983]. *Philosophical Investigations*. Oxford: B. Blackwell.

ESSAY 6

In or *About* Mathematics?
Concerning Some Versions of Mathematical Naturalism

José Ferreirós

In the last few decades, naturalized epistemology and philosophy of science have made great strides. While the word "*naturalism*" sounds like anathema to many a philosopher, almost everyone (it seems) in philosophy of science wants to claim that his or her position is naturalistic. Predictably, this has led to strong ambiguities in the meaning of the term, to such a point that it may even seem advisable to avoid its use. Restricting our attention to the field of mathematics, we still find a complex gamut of different and even incompatible positions. What follows is an attempt to reflect on this issue, in order to stake out my own approach to the philosophy of mathematical knowledge. For this aim it will prove useful to make a contrast with the forms of naturalism proposed by Quine, Kitcher, and Maddy. Much of the discussion will also be concerned with the notion of mathematical practices, again prominent in the agenda of philosophers of mathematics today, the reason being that it is integral to those versions of naturalism that Kitcher and Maddy have proposed.

If anything is clear about naturalism in philosophy of science, this is the fact that naturalists reject *foundationalism*—the view that philosophy has to play the role of a founding discipline, so that work in the sciences somehow presupposes that philosophy has laid the pillars of the building (or at least can reconstruct the pillars, say, by the use of transcendental method à la Kant). This view is often called *philosophy-first*, in reminiscence of Descartes' "*prima philosophia*," which certainly was one of the most forceful and influential foundationalist calls. According to naturalists, Descartes is dead. Philosophy of science is a reflection on the "*factum*" of the sciences, an attempt to understand how scientific knowledge is actually (not 'transcendentally' as with Kant) possible, an attempt to account for scientific knowledge, or even an attempt to explain it. Since this is not a foundational enterprise, the philosopher's account may very well rely on scientific results, on empirical evidence.

1 Two Viewpoints: Kitcher and Maddy

Within the English-speaking tradition, naturalism about mathematics began already in [1969] with Quine's famous paper "Epistemology naturalized." However, Quine's approach contained a deep internal tension, since his naturalistic ideas were elaborated on top of a staunch empiricism. Although Quine failed to realize it, blinded by the tradition in which he had been raised, empiricism is again a foundationalist position, and thus it conflicts with the spirit of the enterprise. The later development of Kitcher's views was, among other things, a process of gradual realization of the need to abandon empiricism. Outside the Anglo-Saxon world, there existed other approaches that we may well describe as naturalistic, e.g., Jean Piaget's views on the epistemology of mathematics (see the short exposition in Piaget [1970]). His ideas are particularly interesting for contrast with Quine, since Piaget was far from empiricism and at the same time he was one of the pioneers of cognitive science.

The views advanced by Philip Kitcher in his [1984], and Penelope Maddy in her [1997], are without doubt among the most successful naturalistic proposals since Quine. As we shall see, they are good examples of the polysemy of the label we are discussing. For Kitcher, naturalism is a view concerning mathematical knowledge, it's naturalism *about* mathematics. Maddy, by contrast, advocates naturalism *in* mathematics. What is meant by these expressions shall become clear in the remainder of this section and the following one.

1.1 GENERALITIES ABOUT KITCHER'S APPROACH

In [1984] Philip Kitcher published an important book, *The nature of mathematical knowledge*, which contained a strong attack against apriorism about mathematics and started the elaboration of his own naturalism. Without entering into the debate of Kitcher's arguments against apriorism, let us focus upon his more positive proposals. These were laid out in two chapters devoted to basic knowledge of mathematics, and three chapters devoted to mathematical change.

At this point in his career, Kitcher tried to elaborate his original ideas and start a new line of development while playing heed to some traditional lines of argument. His new epistemological views were still labelled "a defensible empiricism," and (in agreement with the explication of knowledge as justified true belief) they depended crucially on an account of ontological matters that could make the statements of mathematics true (if only vacuously, see Kitcher [1984]: §6). While this certainly helped him put the philosophy of mathematics in line with general epistemology and philosophy of science, it was also the source of some of the book's weaknesses. Reading it, one has the impression that the author's ideas about mathematics are better than the concrete form in which he presents them: A more rupturist epistemological position would have helped him formulate the key insights more convincingly.

Kitcher's basic position was to analyze current mathematics as a *practice* whose justification, partly historical and partly empirical, is the following. Current mathematical practice has formed as the result of changes in previous mathematical practices, it is the last in a sequence of interrelated practices. The transition from one practice to the next can be seen, in retrospect, to be *rational,* i.e., to agree with one of several patterns of "rational transition" that Kitcher ([1984]: §9) goes on to specify. And the series of practices is ultimately grounded in rudimentary mathematical knowledge, basic knowledge, e.g., of numbers, gained empirically by our pre-historic ancestors. Here he agrees to a large extent with J.S. Mill, and to some extent also with Piaget:

> [Such basic knowledge is anchored in] perceptual experiences in situations where [people] manipulate their environments (for example, by shuffling small groups of objects). What naturalism has to show is that contemporary mathematical knowledge results from this primitive state through a sequence of rational transitions. (Kitcher [1988]: 299)

As one can realize, Kitcher's general strategy was to emphasize the similarities between mathematical knowledge and science in general, and he did so to a point that we can even find exaggerated (I do).

Another aspect of Kitcher's proposal that betrays its origins in Anglo-American philosophy of science of the 1970s, is the concrete explication he gives of the notion of a *mathematical practice* (Kitcher [1984]: §7, in particular 163–4). This is conceived, in agreement with the "linguistic turn," as a logico-linguistic structure, which is framed so as to implement some of Kuhn's insights about normal science. A mathematical practice is a quintuple $\langle L, M, Q, R, S \rangle$ formed by a certain language L, a set of accepted questions Q, a set of accepted reasonings R, a set of accepted statements S, and also a set of "metamathematical" views M (which includes "standards of proof and definition and claims about the scope and structure of mathematics"). Changes in one or more of these sets of linguistic entities will be the motor that drives interpractice transitions, by making instable the previous practice and thus creating the conditions for the emergence of novelties.

It was certainly one of the virtues of Kitcher's proposal that he offered a concrete definition of mathematical practices, where most philosophers of mathematics have tended to be imprecise or even ambiguous. Nevertheless, one may still fear that his definition is too narrow, and that for this reason he was risking to lose sight of some of the richer and deeper sources of the specificity of mathematical knowledge. I return to this problem below.

A few years later, Kitcher acknowledged that the label of empiricism he had been using was "a philosophical fetish" (Kitcher [1988]: 221), opting for "naturalism" instead. But, as we shall see, one can still criticize his account at several levels for agreeing too much with empiricism. We come back to concrete aspects of Kitcher's views later on.

1.2 GENERALITIES ABOUT MADDY'S APPROACH

The evolution of Penelope Maddy presents us with a little bit of a paradox: Having developed a naturalistic account of mathematical knowledge in the book *Realism in Mathematics* ([1990]), she turned away from it in her later *Naturalism in Mathematics* ([1997]). There are of course continuities between both books, especially the crucial question of the introduction of new axioms in mathematics (mainly set theory) and the elaboration of philosophical positions inspired by Quine and Gödel. Maddy's basic starting point was a Quinean conception of mathematical ontology (compromise platonism) and epistemology (continuity between science and mathematics, emphasis on psychology). In her [1997], however, she turned away from her previous attempt to ground mathematical intuition in neuro-cognitive mechanisms, de-emphasizing the continuity between mathematics and scientific knowledge.

This last move was actually motivated by attention to current practice among researchers in set theory and foundations. Noticing that they act as an autonomous community, guided by criteria that are largely independent from the considerations that motivate natural scientists, Maddy was led to argue for a peculiar form of "naturalism." Its main goal is not to understand mathematical knowledge within a larger spectrum of knowledge, but rather to develop a careful analysis of the methods and criteria that guide the development of mathematics (in this case set theory) at the research frontier. Prominent among the questions asked by Maddy is, what are the criteria that guide set theorists in their rejection of some axioms (e.g., Gödel's axiom of constructibility) and adoption of others (e.g., large cardinal axioms)? As I have put it above, the goal is not to elaborate naturalistic views *about* mathematics, but rather to analyze and explicate naturalism *in* mathematics.

The call to focus on mathematical practice can be followed in divergent ways, with different effects as to the extent to which the resulting philosophy will depart from the foundationalist views that dominated the 20th century. As everyone knows, we may follow the call by concentrating on past practices (which one can study rather densely, thanks to the richness of primary and secondary sources available) or on present-day or recent practices (which eliminates the distorting effects of hindsight, and brings us close to the questions that worry today's practitioners, at the price of characteristic difficulties). Along another dimension, we may stay close to the sub-discipline of mathematical logic and foundations (as Maddy does), or we may follow another usual call, the call for attention to a wider spectrum of knowledge and practices which are essential to the discipline, but whose traits remain out of sight for those who concentrate excessively on foundations.[1]

Confronted with all of those options, the reasonable path is to choose carefully one's case studies as a function of the questions one wants to study. And it is also advisable to choose carefully the labels that one uses to describe the orientation being proposed.

Maddy recommended to turn away from traditional philosophical questions about mathematics, since these questions are of little interest to its practitioners:[2] Methodological issues are the main concern of her naturalism. It is very far from my intention to deny the importance of such methodological issues; I fully agree with Maddy on the importance of considering them. Thus it is also far from my intention to deny the interest of her work. What I shall question is the idea that Maddy's "naturalism" is a reasonably complete proposal for the philosophy of mathematics.[3] And I will suggest that Maddy's option for the label "naturalism" was ill-advised.

---

[1] The collective volume Ferreirós and Gray [2006] is an example of the choices opposite to Maddy's—for older practices and away from foundations. A key case of mathematical knowledge that lays claims to specifity, and which was largely forgotten by 20th century philosophers, is ancient and early modern geometry (see Manders [2008]).

[2] For an exactly opposite estimation of the interests of the practitioners, see Ferreirós and Gray [2006]. Of course, any such estimate depends critically on the sample of practitioners one considers, and that is the case there.

[3] This seems to have been Maddy's own conclusion, and in fact she now prefers to call her approach or piecemeal method of inquiry 'second philosophy,' which avoids most of the perplexities that I express in this paper. See Maddy [2007]: 403.

2 Is There a 'Natural' Form of Naturalism?

The bewildering variety of diverging viewpoints, all of them covered by the increasingly broad and vague term "naturalism," is nowadays a well-known phenomenon in the philosophy of science. Philosophers are guided by vogue, and this is the price of vogues. In fact, the variety is much greater than the previous examples may have suggested, ranging from strong sociological programs such as Bloor's, to the most scientistic and reductionistic proposals.

In Quine's original version, the idea was to turn to the sciences (for him above all physics and psychology, again showing the strong effects of empiricism) as the most reasonable way to try to solve the very difficult problem of understanding scientific knowledge, once foundationalism has been fully rejected. Recent authors such as Dehaene ([1997]) and others (see Campbell [2004]) have turned to neurobiology, in a move that may be regarded as beneficial both philosophically and methodologically. However, with Bloor and other sociologists (e.g. Pickering [1995]) we turn away from the natural sciences, and even from psychology, to place all the emphasis on social factors. Is it still advisable to call such an approach "naturalistic"? To make the obvious pun, why not "socialism"? Of course the rejection of foundationalism and the turn toward empirical results are still there, but one clearly feels the need for finer distinctions.

With Maddy, finally, "naturalism" means to turn one's back to *both* the social and the natural sciences, abandoning most of the traditional philosophical questions, in order to concentrate on mathematics itself, on methodological issues and on the criteria at play in decision making within the mathematical community.

Faced with this situation, and knowing the strong appeal of vogues, I have long been inclined to stop using the label "naturalism" altogether. It has been used and abused to the point of making it almost meaningless. Can we still hope for its future? Here I shall make an exercise of optimism and hope that, indeed, we can still correct the ship's course.

Instead of producing a long and careful discussion, I shall allow myself the freedom to be brief. To keep the idea of naturalism minimally close to its Quinean origins and later development at the hands of philosophers of science, I surmise that the following condition should hold:

(1) A naturalistic philosophy of X shall include an account of *knowledge* in the field of X that is non-foundationalist, and that relies on scientific results; at the very least, such an account should aim at establishing contact with results of the sciences (possibly in the form of a future convergence).

But, given current knowledge of human knowledge (not only from a scientific perspective but also from a number of philosophical perspectives), it is natural to assume that

(2) A naturalistic theory of knowledge in the field of X shall rely on results from the *natural* sciences—it shall conceive of knowledge as produced by communities of living beings (the members of our species) which not only engage in interactions among themselves, mainly through language and images, but also in interactions with their physical environment.

Reference to the natural sciences in general, and the biological grounding in particular, is in my view essential. But notice that this does not imply to take a reductionistic approach.

If we incorporate the above two conditions into our talk of *naturalism,* it is obvious that Maddy's approach does not qualify, nor does Bloor's. But it should be obvious that this is not meant to deny the interest of their work: I am only questioning the appropriateness of using the label "naturalism" for it. It seems to me that much of Maddy's work in [1997] is not at all incompatible with a naturalistic philosophy of mathematics in the sense just delimited.

I believe it is advantageous to emphasize biology to the detriment of psychology, since work in the latter field encourages all too often the persistence of what Dewey called "spectator theories of knowledge," towards which I am (like other philosophers and at least some scientists) quite sceptical. That's why I said that it is beneficial philosophically, since it distances us from mentalism and excessive emphasis on the so-called internal world. Reference to biology is also most promising (in my opinion) given the current state of research, especially on the experimental side. Here lies the methodological appeal of this move, since advances in the available experimental methods have been quite impressive (e.g., neuroimagery by magnetic resonance).

But the view that an appeal to the natural sciences should figure prominently in any sensible kind of naturalism is by no means incompatible with appeals to the social. The fact that science and mathematics are produced by *communities* of humans that exchange information through language, writings and drawings (or other images), does not conflict with the claim that the members of those communities share a biological constitution and a physical environment. But although the idea is obvious enough when clearly stated, in practice many authors argue as if that was not the case.

Aristotle chose to conceive of humans as the "zôon logon ejon (ζωου λογου εχου)," the *language animal*[4]. That conception, properly understood and developed, points the way to an integration of the natural-science and the social-science aspects of human behavior and human cognitive abilities—including the abilities that make mathematical knowledge possible.

What has been said up to now is by no means enough. Very often, one finds that talk of naturalism is a way of keeping alive, *aggiornati*, old scientistic beliefs that leave no room for philosophy as a form of critical reflection. My own understanding is that naturalism does not imply scientism, and that it is quite essential to preserve forms of critical reflection: The philosophy of mathematics ought not to be simply a field for the naïve "application" of some theories taken from the cognitive or the natural sciences.

Often, too, the views that some proponents of naturalism are exploring involve naive or crude forms of reductionism. Obviously enough, the most popular brands nowadays are brain reductionism and gene reductionism: conceptions that knowledge should be conceived as "nothing but" activation patterns of neural networks in the brain, or even that the key to understanding, e.g., language is to find the "corresponding" gene. Such assumptions are more than questionable, not only on philosophi-

---

[4] In my opinion, a better translation than the traditional "rational," too charged with extra connotations, too suggestive of dichotomies that a naturalist aims to supersede.

cal, but on purely scientific grounds. Hence I consider it undesirable to include them as characteristic marks of naturalism. If this were to be the case, I should certainly describe my own position as non-foundationalist and non-naturalist. It is undesirable that the label naturalism may serve to hide differences as to such assumptions.

The reader should not interpret my criticism of naive reductions as some kind of appeal to mysteries or the supernatural. The claim is just as mysterious as the idea that the genome by itself, in the absence of adequate surrounding conditions such as are found inside the cell, is not functional. Concerning this point, let me add the following. Sometimes the enemies of naturalism show great interest in crudely reductionistic versions of this position, since they can argue their inadequacy relatively easily, and from there they argue the need for a position that transcends naturalism. But such an argument, as it should be obvious by now, is a *non sequitur*: It is based on a false dichotomy of the form "either crude reductionism, or non-naturalism."

3 Some Presuppositions: "Formal" Sciences and Hypotheses in Mathematics

In the paper "Epistemology naturalized" that was mentioned above, Quine elaborated on his principled rejection of the dichotomy formal/empirical sciences (based on his well-known holism, which emphasized the continuity between science and mathematics, and his arguments against the distinction between analytic and empirical statements). Kitcher has basically followed his lead in conceiving of mathematics as continuous with the empirical sciences. Maddy, on the other hand, has departed more and more from Quine's standpoint, being finally driven to an emphasis on mathematics as a fully autonomous discipline.

The question can hardly be avoided, as it will inform (explicitly or implicitly) any positive views we may propose concerning the nature of mathematical knowledge. Once again, in the sake of brevity, I will allow myself the freedom to avoid detailed arguments. The distinction between "formal" and "empirical" sciences is in my view ill-posed and misleading. It would not be so, if one could believe that the statements of mathematics are mere tautologies, or are based on purely intra-linguistic conventions, so that their contribution to the meaning of scientific claims about empirical states of affairs is null. Such were the views of Wittgenstein in his early work, and Carnap throughout his life, but not our views.

The point is not so much to deny the peculiar nature of mathematical knowledge. Although we certainly wish to claim that mathematics does *not* belong "in a totally different sphere" from that where the sciences belong, one may still argue that there is a significant difference in the nature of the cognoscitive enterprise that we call mathematics, compared with the observational and experimental sciences (be they "social" or "natural"). The point is, above all, that by trying to capture this difference by means of the labels "formal" and "empirical" we are severely misrepresenting its nature. And here I am not primarily thinking about methodological issues such as the presence of quasi-empirical or experimental methods in the actual practice of mathematics. The existence of such methods is certainly relevant, but there is more.

The basic principles, the stipulations on which mathematics is based, are not purely formal; and, when such principles are employed as a basis for laying out models of natural phenomena, the mathematical stipulations *do contribute* to the empirical claim being made. Whether or not Quine was right in his arguments against the distinction analytic/empirical, I believe he was very right in arguing that it is impossible to adjudicate the so-called "empirical content" of scientific statements in such a way that some such statements distill all that is empirical, and some others remain totally disjunct from the first kind.

In fact, here—as with some other of his characteristic theses—Quine was not entirely original. Earlier authors, particularly Riemann and Poincaré, Hilbert and Weyl, had made the very same point. I quote Weyl comenting on Hilbert's foundational work:

> Hilbert furthermore pointed with emphasis to the related science of *theoretical physics.* Its individual assumptions and laws have no meaning that can immediately be realized in intuition; in principle, it is not the propositions of physics taken in isolation, but only the theoretical system as a whole, that can be confronted with experience. What is achieved here is not a perceptual insight into particu-

> lar or general states of affairs or a *description* that faithfully copies the given, but a theoretical, in the last analysis purely symbolic, *construction* of the world. [Weyl goes on to explain a Machian distinction between "real propositions" of physics, which establish pointer coincidences, and "ideal assumptions" of which no perception directly gives the full meaning; and remarks that our theoretical interest attaches very especially to the ideal assumptions.] . . . According to Hilbert, already pure mathematics goes beyond the bounds of intuitively ascertainable states of affairs through such ideal assumptions.
>
> What "truth" or objectivity can be ascribed to this theoretic construction of the world, which presses far beyond the given, is a profound philosophical question.
>
> (Weyl [1927]: 484)

Idealisation via the elaboration of scaffoldings of concepts (Hilbert's *Fachwerke von Begriffen*) and the *introduction of hypotheses* are essential aspects of mathematical activity.

As I have defended elsewhere, the hypothetical nature of some mathematical assumptions (basic principles or axioms of advanced mathematics) is a key realization that emerged from the foundational debates of the early 20th century. The deeply consolidated divide between *constructivist* mathematicians and "modern" or *postulationist* mathematicians attests to this. Examples of relevant *hypotheses* are: the axiom of choice and several other axioms of set theory (infinity, power set, etc.); even more obviously, the basic assumptions of geometry; and, perhaps less obviously, the axiom of completeness for the set of real numbers.[5]

In my use of this word, "hypothesis," the usual connotation of representing something in the so-called "external world" is excluded, and I do not intend to imply that they receive inductive support from some "basic" statements. (Even much less do I mean to establish—like Quine or Gödel—some kind of comparison between postulates concerning sets and the assumption that there are physical objects.) The word is being

---

[5] For further details see Ferreirós [2005a] and the Introduction to Ferreirós and Gray [2006]. This topic has been important for many philosophers of mathematics, including Lakatos ([1967]) and Maddy.

used in its etymological sense, to denote that which "lies behind" and supports a certain theoretical building, a certain scaffolding of concepts and principles. Indeed, Aristotle used the word "hypothesis" ($\dot{\upsilon}\pi\acute{o}\theta\epsilon\sigma\iota\varsigma$) to refer to what we call axioms, and Riemann adopted this very same usage again in 1854, exploiting the intervening changes in the word's connotation. Here I follow both, and by doing so I mean to emphasize (like Riemann [1868]) that some mathematical axioms are not *a priori* or endowed with certainty in any way, but rather, they are bold theoretical stipulations. And I believe, again like Riemann, that the assumption of the *continuum* (the continuous structure of Euclidean space, the topological structure of the real numbers) is the most crucial one, both conceptually and historically, as it has induced the introduction of several other assumptions into modern mathematics.

Such a claim does not force us to the conclusion that *all* of mathematics is purely hypothetical. Telegraphically, let me say that I believe we have *certainty in our knowledge* of the theory of natural numbers, and that knowledge justifies adoption of the seven axioms of Dedekind–Peano Arithmetic. Our certainty concerning claims about the natural numbers (based on our informal grasp of the theory, not on formal-syntactic considerations) is likely to be the ultimate source of our firm conviction that the formal theory is consistent. (Sources of this kind are, in my view, of greater cognitive import than technical results such as the relative consistency of PA with respect to intuitionistic arithmetic.[6])

On the other hand, the assumption of a *set of all* natural numbers is nothing but a hypothesis (the axiom of infinity). As the reader will realize, one of the most interesting questions that emerge within the context of this analysis of mathematical knowledge, is the "dialectics" of hypothesis and certainty. Some aspects of this can be intricate, as witnessed by the case of geometry.[7]

---

[6]Similarly for Gentzen's results. The significance of proof-theoretic results in the tradition of Gentzen would lie more in the direction of comparing the consistency strength of different theories.

[7]There is good reason to believe there is a "core of certainty" in elementary geometry (cf. the seeming universality of some geometrical concepts, or the extent to which there exists agreement between Greek geometry and, e.g., Chinese results about right-angled triangles), but if I am right in the claim made above, that the continuum structure is an idealized hypothetical assumption, it follows that not all

Although some or even many authors have presented the concept of set as an idea that has deep intuitive underpinnings, in my opinion this is not correct. It is incorrect both as a cognitive claim concerning the process by which we learn set theory, and as a historical claim about the process through which set theory was formed in the period 1850–1940.[8] Kitcher ([1984]) proposed an ontology for mathematics that made the concept of set appear to be a reflection of a basic "mathematical reality," which in the end was a description of super-human operations performed by ideal beings. Borges would have reminded him that even an "immortal man," if he were to "exhaust his eternities counting," would never reach the "vast numbers" of Cantorian set theory (see the full quotation in Ferreirós [1999]: xi). In my opinion, that was a wrong move of Kitcher's, and the concept of set can only be presented as based on hypotheses that mathematicians have introduced relatively recently (see below). That concept and these hypotheses are peculiar, because they are very densely connected with all kinds of other elements in mathematics, including the notoriously important idea of the continuum.

I am of course not the first to realize that theoretical elements play such an important part in fixing the concept of set. The usual tendency to trivialize it somewhat, presenting that concept simply as an elaboration of the commonsense idea of a collection, contrasts with the tendency of more philosophical accounts to emphasize the historical motivation coming from the background of 19th century analysis. This in my view offers exactly the opposite message, but many people seem not to realize it, so that what is done in practice does not go into conscious realization.

The process of fixing a concept of set is a theoretical one, and in the last analysis it is solved by establishing an axiom system. (You may follow Hilbert and call the axioms "defining conditions of the concept" if you wish, but that doesn't change anything.) Intuitive pictures or im-

---

of Euclidean geometry is endowed with certainty. Where lies the boundary line? And, which cognitive processes can explain our knowledge of the certain parts of geometry? Such questions seem highly non-trivial.

[8] See the new Epilogue to the 2007 paperback edition of my [1999] book. There I argue that even the view that set theory was the outcome of articulating and making compatible two different "intuitions"—the logical idea of a class and the mathematical idea of a quasi-combinatorial set—does not stand the test of history.

ages, and commonsense conceptions, are useful here and there to help in the process, but they are like a ladder that one must throw away after having used it. In the usual process of climbing (say in an introductory course), certain explanations and proofs, and a good number of exercises, are essential: They provide *fine tuning* of our understanding of the concepts, forcing us to *correct* the rough initial images (e.g., to move away from the commonsense idea of a collection). From a cognitive viewpoint, one ought to say that the concept of set is not constructed bottom-up, from everyday notions and situations, but *through the interaction between knowledge and images at the bottom and theoretical ingredients coming from the top*. If we look at it from a non-introductory perspective, the axioms do the essential work—although their interplay with previous knowledge is likely to be indispensable (see below).

4 On Rudimentary Knowledge of Mathematics: Beyond Empiricism

As we have seen, Kitcher proposed some ideas about rudimentary knowledge, according to which mathematics is ultimately based on perceptual knowledge that was gained by our prehistoric ancestors. People performed operations and manipulated their environments, for example by grouping together a small collection of objects (perhaps in different ways and orders, but always coming to the same end-result). By doing so, they obtained reliable knowledge about rudimentary, practical counterparts of the properties we call commutativity and associativity, about the invariance of the result of grouping (and counting) regardless of differences in the way the objects being counted are grouped, etc. Subsequently they used such knowledge in their counting practices, for instance when they marked tallies on a bone in order to find out how many sheep they had (perhaps to check whether one might be lost upon coming back home at dusk), or to find out how many nights went by from new moon to new moon (say with the Lebombo bone, ca. 35,000 years old).

Two aspects of Kitcher's account have always surprised me: his emphasis on *perception*, to the point of hiding away manipulation or motor action; and his emphasis on the idea that such empirical knowledge was gained by our ancestors in *prehistory*. I believe that only a strongly empiricistic context can explain both traits. They contrast rather sharply with, e.g., Piaget's theories, which from the very beginning emphasize what he called the "sensory-motor" system (perception plus motor action), and the idea that each and every child, prehistoric, or modern, performs the learning process.

The empirical processes that we are talking about are primarily processes of human concrete action, of manipulation, and even technical practices (as in the case of counting with the Lebombo bone); of course, they involve perception, but there is more. However, the mentalism of Western philosophical traditions (empiricism in particular) promotes a reconception of any such situations in terms of pure perception or even sensation (as with Quine). Instead of discussing directly the manipulation of objects, Kitcher ([1984]: §5 and 6) put emphasis on perceptual knowledge, hence on the "perception of" objects being manipulated.

Instead of discussing directly the manipulation of objects, Kitcher ([1984]) talked about "the perception of" objects being manipulated.

To follow such philosophical tendencies is, in my opinion, quite wrong: It seems likely that the worst mistake of Occidental theories of knowledge was, precisely, the *oblivion of action,* its elimination from accounts of knowledge.[9] One of the advantages of naturalism is that it frees us to include into our accounts of knowledge physical objects, human bodies, and concrete motor action, without having to restrict ourselves in a contrived way to perception and language.

The cognitive process of perception is best conceived as a complex high-level process that is *conformed by the feed-back* between sensory inputs *and* motor outputs. The fact that changes affecting the fit between sensory input and motor output do have a very strong effect on perception is well established. A clear example is the well-known experiment in which a subject wears glasses that invert the image upside-down on the retina, a situation in which the perceptual system readjusts the im-

---

[9] That's often based on the assumption that, in order to become a philosopher, one has to follow Descartes and learn to conceive of the world as a matter of perceptions in a mind. It should suffice to evoke the names of Dewey, Wittgenstein, and other philosophers (including Heidegger and Marx), to see that such an assumption is dubious. In order to be a philosopher today, one has to learn the Cartesian move, and then must learn to see that it is questionable.

age and re-establishes the fit after only 24–72 hours.[10] This is very clear evidence for the celebrated plasticity of the brain, and in particular for so-called perceptual recalibration, which puts that plasticity in the service of adapting perceptual images to serve motor action and practical needs.

As regards the period when such empirical knowledge is or was gained, and Kitcher's insistence on prehistory, he was probably concerned to avoid sceptical arguments claiming that people like us have not obtained such basic knowledge directly, but rather "inherited" it with the language we have learnt. In taking this course, traditional philosophical concerns once again occupied too much space in the naturalist's agenda, leaving important things out. It is not my intention to diminish the importance of questions about the role of language in thought, and whether (as some have claimed) there exist forms of pre-linguistic thought. Such questions have to be asked and carefully investigated, but it seems wrong to avoid making important points out of fear that one extreme position in that debate (in our case, absolute denial of pre-linguistic thought) may be right.

It seems perfectly safe, indeed grounded in experience, to assume that rudimentary knowledge based on concrete operations and manipulation is gained once and again by members of each human generation. To continue with Kitcher's example (groupings of objects and their properties), children obtain knowledge about them directly, but also and very especially by learning to memorize the numerals and to employ them for counting. This phenomenon is well worth detailed consideration, quite independently from more ambitious answers to the question of language and thought.

Numerals have a very special role within language and vocabulary, so much so that numerals have been claimed to be "to an extent self-contained," having "distinct characteristics" to the point that in some respects they are "unlike almost anything else in language."[11] Linguists such as J. Hurford claim that the numeral system of a language is much less susceptible to change than most other parts or systems (unless the

---

[10] G. Stratton was the first to perform such experiments in the 1890s, though this line of research links back to Helmholtz (see Dolezal [1982] and Kohler [1964]).

[11] See Hurford [1987] for the linguistic evidence and an assessment.

numeral system is replaced by that of another dominant language, as happened to the Japanese and Thai numerals, replaced by the Chinese). This rather unique position of numerals can be explained by the fact that they are a very particular kind of linguistic instrument, employed in a most distinctive type of speech act. Indeed, counting can be regarded as nothing but a *technical practice* of mankind, one of the oldest, and one of the earliest we consistently teach our children, generation after generation.

In short summary, counting consists in:

(1) "conceptually" considering a collection or aggregate of objects (by a collection I mean a class-as-many, not anything like a set);[12]

(2) producing a finite number-sequence via number-words or number-symbols, or even a sequence of strokes (like the primitive shepherd or the Lebombo woman); and

(3) effectively pairing them up by means of a concrete (finite) one-to-one correspondence.[13]

Learning the proper use of the numerals in counting is quite independent from other aspects of language-learning, just like the kind of things we do with the ordinals or counting-words is very different from the acts performed with other parts of language. I mean to suggest that the exploration of possibilities of operation upon our immediate environment, which is made possible by learning to use the numerals, is like the possibilities that were open to our prehistoric ancestors by using tallies on a bone. There is nothing specifically linguistic in this process, that would place it in a separate category (say, of linguistic organization of an otherwise disorganized or chaotic environment).

---

[12] I choose to speak in Frege's parlance, even though the question seems to me more complex than he assumed: One should take into account critiques of the idealized conception of "concepts."

[13] Dedekind's [1888] analysis of number departed from these simple observations, but framing them within an abstract setting: He replaced collections (in the above sense) by sets; the constructed number-sequence by the number structure; and the concrete, effectively-given correspondences by mappings.

Operations and practices having to do with grouping and counting by no means exhaust the phenomena that are relevant to a discussion of elementary mathematical knowledge. Other important practices, which to be sure are much more complex, but which again could be called "technical practices," have to do with figures (e.g., the drawing of circles and right-angled triangles, practical geometry) and with measuring by rule or with measuring units such as "casks" or "barrels" (which leads e.g. to the introduction of fractions). However, this is not the place to enter into a more detailed discussion of any of these.

On a different line, the question of rudimentary knowledge of mathematics is one about which there has been some progress in recent years. It has long been known that there exists a specific neural ability of "subitizing," i.e., grasping immediately very small numbers of similar elements. This has been observed in infants, in other mammals such as mice, in pigeons, and according to some experiments even in newborns. For humans, the limit of our ability to grasp the numerosity of some elements directly lies around number 5. In recent years, this phenomenon has been studied by means of neuroimaging techniques (Dehaene [1997]), and it has been linked with a seemingly innate capacity to estimate larger numbers, which Dehaene explains by appeal to an analogical "accumulator."[14] Such studies indicate that there are specific cognitive abilities underlying our grasp of number, so that the innate ability to grasp very small numerosities seems to lie behind our capacity to use and understand the first few numerals. All of this of course is still far from the *concept* of number, which arguably should only be regarded as present when there exist specific techniques (normally involving language, number-words, or number-symbols, though in prehistory tallying would also do) allowing the process of counting and its extension to larger numbers. Typically these practices always combine the cardinal and ordinal aspects of numeration and involve concrete, effective one-to-one correspondences.

---

[14]Corresponding studies about our grasp of figures are still a task for the future, but some interesting anthropological results are already available (Pica and Dehaene [2006]).

5   On Mathematical Practice: Beyond Kitcher's Kuhnian Picture

As I remarked above, one of the virtues of Kitcher's proposal is that he linked his naturalism with the idea of mathematical practice and offered a concrete definition of mathematical practices. His approach was heavily influenced by the then-in-vogue models of scientific change, and by the preference of analytic philosophers for linguistic formulations. A Kitcherian mathematical practice is a quintuple $\langle L, M, Q, R, S \rangle$ formed by a certain language L, a set of accepted questions Q, a set of accepted reasonings R, a set of accepted statements S, and a set of "metamathematical views" M. Since the proposal is to reconstruct the development of mathematical knowledge as a sequence of practices linked by rational transitions, Kitcher's approach includes the suggestion that mathematics at any given historical period is governed by one single quintuple $\langle L, M, Q, R, S \rangle$. As remarked above, this is very close to Kuhn's notion of normal science.

The appeal to mathematical practice is very suggestive, but in principle it is merely a general recommendation to look closely into whatever the members of a mathematical community are doing. Included would be their problem-solving practices in different contexts, their practices of drawing and deploying pictures, their proof practices, their standards and criteria of evaluation, etc., but also practices of writing and communicating, not excluding teaching practices. And of course a sociologist or anthropologist would not stop there, but insist that one must develop social research on scientific practices in a non-biased way, considering the practices of institution and community building, and following mathematicians wherever they go in society.

Compared with that broad spectrum of possibilities, Kitcher's notion of a practice is very narrow and strictly regimented, as fits a follower of Quine. But it is not so strict as it might seem in light of its superficial similarity with the definitions of logicians and "semantic" philosophers of science (from Suppes and Sneed to van Fraassen). The five elements L, M, Q, R, and S are not conceived to be formalized, nor are they mathematical structures. For an interesting example, one might begin to research what those five elements would be in the case of Cartesian analytic geometry: The language L, for instance, is partly symbolic and partly a refined natural language, and perhaps it has to include the drawings of geometric figures, since they are still essential in the conduct of proofs. And what about the metamathematical views? This again is a rather complex matter, for those views are quite unlike the metamathematical results of 20th century mathematical logic: They include ideas about the generality and heuristic power, but lack of rigor, of algebraic manipulations, contrasted with the rigor but insufficient generality of geometric proofs, and a long etcetera, which does not stop short of entering into the realms of metaphysics and theology.

5.1 BEYOND KITCHER'S PRACTICES

There are several possible reasons to fear that Kitcher's definition may be too narrow. This happens even if we restrict our attention to what might be called the "cognitive" practices of mathematicians (thus leaving aside sociological issues). First, there is the general problem of how best to approach the question of practices, and some authors are inclined to believe that an overarching concept in the style of Kitcher is less useful than the idea of a diversity of topic-specific practices in coexistence and interaction. Second, there is the question whether Kitcher's adoption and adaptation of the Kuhnian idea of normal science is faithful to the historical reality of mathematics. In recent years, historians have emphasized the role of diverging traditions, differently-oriented research schools, and their interactions, within a single historical period. Third, there is the complex question whether by imposing a linguistic approach we may not be losing sight of some of the richer and deeper sources of the specificity of mathematical knowledge.

Let me say a bit more about the last two problems, beginning with the third. Speaking very generally, a naturalist must consider seriously the role of non-linguistic cognitive elements in the conformation of mathematical knowledge. That they do play a role seems to be the case with elementary geometry, not only when we turn our attention to cut-and-paste practices (such as those of Bhaskara in 12th century India), but also in the celebrated and paradigmatic case of Euclid. In Euclidean geometric practice, even if we restrict our attention to what is essential to the conduct of proofs, the diagrams and some specific ways of using them are at the core.[15] It can even be argued that the Euclidean axioms are nothing but the basic stipulations that regulate the construction of diagrams (which explains the original form of the parallel axiom, since it specifies conditions under which the vertex of a triangle can be found), and that underscores the enormous distance between Euclid's axiomatics and the modern Hilbertian form: They emerge as almost totally different businesses. For another example, we have previously discussed evidence for the role of non-linguistic elements such as subitizing (see above) and concrete one-to-one correspondences in connection with the roots of the number concept. The same is most likely to be the case in relation with measuring practices, which are at the source of the introduction of fractions and proportions (rational and real numbers in the modern conception).

Of course, many authors have been inclined to disregard all of those elements as belonging merely to the "context of discovery" of mathematical systems. These systems would be purely symbolic, preferentially axiomatized in the style of Hilbert, or even formalized in accordance with the practices of 20th century mathematical logic. Without entering into a discussion of this alternative view, or even pausing to spell out the way in which it annihilates the history of mathematics up to 1900, it suffices here to say that the approach I will be proposing is based on exactly the opposite assumption. At this point, I am content with the mere claim that such an approach should be regarded worth exploring—at least by those who favor naturalism!

Coming back to the second problem, monolithic unity à la Kuhn vs. diversity, there exist paradigmatic examples in the history of mathematics, and the more carefully we study that history, the more they tend to abound. One of them is well-known to experts in foundational philosophy of mathematics, namely the contrast (and coexistence) between the traditions of "modern" postulational mathematics and constructivistic mathematics. But there is much more. Take for instance a crucial part

---

[15]See Manders [2008], in particular his ideas about exact and co-exact determination of elements in diagrams, and about the role of objectors in diagrammatic practice.

of the development of mathematical analysis in the period 1860 to 1914, the divergence between the Weierstrass and Riemann schools.[16] This contrast in late 19th century analysis can only be captured by *different* (though coexistent) Kitcherian quintuples: Maybe the set of accepted questions Q, and many of the accepted statements S, are common to both schools; but certainly not so for the accepted patterns of reasoning R, the "metamathematical" views M, and even, as a consequence of those two, the basic language L in which analysis (be it real or complex analysis) was meant to be formulated—what may seem most outrageous, from the standpoint of the analytic tradition.

To explain the contrast a bit, consider the particularly clear case of the opposition between Riemann's and Weierstrass's approach to the theory of analytic functions. Weierstrass systematically preferred explicit representations of analytic (holomorphic) functions by means of power series of the form

$$\sum_{n=0}^{\infty} a_n(z-a)^n,$$

connected with each other by analytic continuation. Riemann chose a very different and more abstract approach, defining a function to be analytic *iff* it satisfies differentiability conditions expressed in the Cauchy–Riemann equations:

$$\frac{\partial u}{\partial x} = \frac{\partial v}{\partial y} \quad \text{and} \quad \frac{\partial u}{\partial y} = -\frac{\partial v}{\partial x}.$$

This neat conceptual definition appeared objectionable to Weierstrass, as the class of differentiable functions had never been carefully characterized (in terms of series representations); exercising his famous critical abilities, Weierstrass offered examples of nowhere-differentiable continuous functions.

Thus, Riemann was favouring a more modern language, which pointed the way towards structural mathematics, while Weierstrass stayed closer to 18th century calculus, e.g., the idea of a function as an analytical expression. (Riemann favoured Dirichlet's idea of a function as any

---

[16]For details see Ferreirós [1999]: §I–II, but especially Bottazzini [1986] and Tappenden [forthcoming].

well-defined but possibly "arbitrary" relation between numbers, not required to be given by an explicit formula.) On the "metamathematical" level, Weierstrass saw the essential point in building up analysis rigorously from arithmetical conceptions, in a way that can be described as semi-constructivistic: This was his version of *arithmetisation,* and that is why he always emphasized series representations. Riemann, meanwhile, was worried about finding the adequate conceptual characterization of the mathematical notions, and subordinated the methods of reasoning to this goal—which is why he opted for methods that were not rigorous at the time, but which pointed the way to modern mathematics (topological ideas and variational principles that would be rigorized in the early 20th century). Riemann found objectionable the redundancy of information about the function given by sequences of power series, and made a conscious effort to characterize the functions by independent traits.

Some readers might be tempted to regard the contrast I have described as an instance of 'revolutionary science,' and thus compatible with a Kuhnian account. However, the historical fact is that both mathematical schools emerged roughly around the same time (in the 1850s) and they developed in parallel during the following half century.[17] Thus the situation cannot be modeled by Kuhn's scheme (normal science / revolution / new period of normal science) nor by means of two Kitcherian practices and the "rational transition" from one to the other. To conclude, this and other examples show that the historical development of mathematical knowledge is, during some periods at least, marked by a plurality of schools, each one of which should have to be described by a different Kitcherian quintuple.

5.2 ON PRACTICES AND THEIR INTERPLAY

Let me come back to the general problem of how to frame and deploy the idea of practices. I shall explore the implications of some basic ideas about practice for our understanding of mathematical knowledge. My approach here will be rather minimalistic, based on some of the *simplest* traits of a practice-oriented account. Even if we disregard subtler

---

[17] For details see Bottazzini [1986] and Tappenden [forthcoming].

aspects of mathematical practices such as the images of mathematics, the values, and so on, we are still left with sufficient material for an interesting analysis of the constitution of mathematical knowledge. This will be shown by focusing on an exemplary case: the way in which previous mathematical practices (in particular arithmetical ones) have constrained the admissible principles and results of set theory.

We shall employ the phrase "mathematical practice" in a way that remains quite close to Kitcher's usage, although I am not sure whether this is the best terminological choice: One might just as well speak of *strata* of mathematical knowledge, instead of practices.

My emphasis is on mathematical practic*es* in the plural, for it is a key thesis of this approach that *several different levels of knowledge and practice are coexistent,* and that *their links and interplay are crucial to mathematics.* This implies that I question the Kuhnian image of "normality" that Kitcher followed, but not (at this level) in the sense of the above remarks about different but coexisting mathematical schools. By coexistence I mean that different practices or strata live together within a single historical period, and furthermore that they live together and interact within the mathematician herself.

To give the easiest example: Any mathematician has knowledge of the practices of counting (gained in early childhood), of basic arithmetic or calculation (gained in primary school), and of the number structure $\langle \mathbb{N}, 0, \sigma \rangle$ (obtained in college; here σ is the successor mapping). These are three different practices, the last two of which can be described (on a first approximation[18]) by means of two different Kitcherian quintuples $\langle L, M, Q, R, S \rangle$; and the three practices coexist. In a textbook or encyclopaedia presentation it makes sense to systematize and present all of these things within a single framework, but that is not a faithful representation of the cognitive structures at work. A naturalistic and practice-oriented study of mathematical knowledge must emphasize the level breaks, and the interrelations. To turn it into a slogan: Systematicity is a trap, not a virtue, for a naturalistic epistemology.[19]

---

[18] As I have suggested above, my own preference would be to broaden Kitcher's concept of a practice and introduce in it elements that are not linguistic, such as drawings or geometric images, and I would even give an important role to exemplars in the sense of Kuhn. But for present purposes we can simplify.

[19] This is yet another point where my theoretical preferences run against Quine's—

Structures (a set-theoretic practice)
$0 \notin \sigma(\mathbb{N})$, with σ the successor function

↑ ↓

Calculation (a symbolic practice)
$3 \cdot (17 + 5) = 51 + 15 = 66$

↑ ↓

Counting (a *technical* practice)
'uno,' 'dos,' 'tres,' 'cuatro,' . . .

Figure 1: A schematic picture of interrelated practices. In each case, only a representative instance is given.

I have said that the links and interplay between different practices or strata are crucial to mathematics. Crucial, that is, for the constitution of meaningful concepts, the determination of admissible principles, and the development of mathematical knowledge through the rise of new practices. In the sequel I shall present some examples, but for now consider again the number concept and its associated practices of counting, calculatory arithmetic, and the number structure $\langle \mathbb{N}, 0, \sigma \rangle$. Part of the constitution of our knowledge of numbers consists in establishing *systematic interconnections* between elements in the three practices: We know for instance how to solve a practical numerical problem, beginning with counting and following with some easy calculations.[20]

Now, it is my working hypothesis that *meaning and conceptual understanding* are gained only when the web of interconnected practices is at play, and in particular when the connections to the more basic

---

see Ferreirós [2005b].

[20] Imagine for instance that I have to count the number of soldiers in a company, and they are on parade, organized by sections: I count the number of rows and columns in a section, and then I switch to calculation and multiply both; again I multiply by the number of sections, which I must have counted before; and maybe, if there are some men or women outside the sections, I may count them and add to find the total.

practices are present. That is how the symbolic systems employed by mathematicians become stably associated with a conceptual content, in such a way that this can be reliably taught and learnt. To put it otherwise, the constitution of meaning in mathematics is intimately related with the interplay of practices, and technical practices play an essential role by anchoring the cognitive roots of the whole system. (I want to emphasize the idea that direct manipulation or intervention in the world is at the roots of mathematics: This is the case with such basic and pervasive practices as counting and measuring.) Imagine that somebody, Paula for instance, learns the mathematical facts about the number structure $\langle \mathbb{N}, 0, \sigma \rangle$ in complete isolation, from a teacher who carefully hides the connections with the more elementary practices. (In the case of the natural number-structure this would be quite difficult, since the connections would easily spring to mind; it may be feasible with the real numbers presented as Dedekind-cuts.) When, after having become very familiar with that mathematical practice, she is finally presented with the hidden connections, Paula would have the typical *aha* experience—the pieces of the puzzle would suddenly be organized, meaning and conceptual understanding would be gained.

In section 3 I remarked that, from a cognitive viewpoint, advanced mathematical hypotheses are not "constructed" bottom-up, from more basic notions and practices, but *through the interaction between knowledge at the bottom and theoretical ingredients coming from the top (axioms)*. That view, coupled with the standpoint on meanings and concepts that I have just sketched, has interesting consequences: We are led to the idea that a mathematical concept is modified when it becomes linked with new practices or strata; in particular, we should consider that the number concept *was modified* when there emerged the idea of a set-theoretic number structure $\langle \mathbb{N}, 0, \sigma \rangle$. And if we bring the same idea to the level of the individual mathematician, the suggestion (which seems to me well worth exploring) is that different mathematicians, with specialised knowledge of different practices, may have *different concepts* of such things as the natural numbers or the real numbers. To put it in a nutshell, a different sub-web of practices involves a different conceptualisation.

Being an approach that emphasizes the links between diverse practices, my perspective is naturally centred upon the *mathematician as an epistemic agent* that establishes those links. As a short comment on this, an agent must have at her disposal cognitive and social abilities in order to become an epistemic agent. Behind the conduct of science as such, there are material conditions that we should take seriously.[21] And it is obvious that knowledge is a social product: Established knowledge can only be gained, and new knowledge can only be established, through immersion in a web of social and intellectual life.[22] Thus, I believe the approach laid out here is perfectly capable of adopting—and balancing—inputs from both the cognitive and social studies of mathematics. But there are also normative conditions that regulate the practices, helping us identify them and also deciding when an agent has sufficient control on actions and their outputs to be considered as an able practitioner. This is what we mean by saying that the approach is centred upon the mathematician as an epistemic agent—not merely a social agent or a cognitive agent.

One of the most characteristic theses of my approach is that the web of practices involved in the constitution of mathematical knowledge includes, not only practices that can properly be called "mathematical," but also other practices. These are typically of two kinds: technical practices, and scientific practices. Considering again the example of the number concept, counting is not mathematics properly speaking, but rather a technique, a very basic technical practice. In counting, as we teach it to little kids, there is a coordination between body movements and perceived phenomena, mainly through the vehicle of oral language, the spoken numerals (see above). The numerals can be regarded as a man-made instrument, so oral language plays the role that sticks, cutters, and tallies play in other counting practices; our hands and body movements also play a crucial role in both cases. As was al-

---

[21] It is well known, e.g., that there exist cognitive impairments that make it impossible for people who suffer them to handle numbers and counting; see Dehaene [1997].

[22] See Wilder [1981]. Notice that immersion in a book is actually immersion in the social exchange of information. Under certain conditions, in the lack of the appropriate social environment (think of hieroglyphic writings before Champollion deciphered the Rosetta Stone), written information can be inaccessible.

ready remarked, other examples of technical practices directly relevant to mathematics are measuring and practical geometry.

Also important are the links with scientific practices, which in turn can be dependent upon mathematical practices. A clear and historically central example is the connections with (what today we describe as) modelling practices in the field of astronomy, say with Eudoxan and Ptolemaic models of planetary motion. Their role in keeping Euclidean geometric practices alive, and their connection with the emergence of trigonometry (or even the notion of function), should be noted. We cannot go into further details here, but as you can see the proposed approach, with its emphasis on the interconnections between mathematical practices and other kinds of practice, ceases to pose the problem of the applicability of mathematics as external to mathematical knowledge itself. It becomes internal, and the very phrase "applicability of mathematics" is turned dubious, based as it is on a puristic (sometimes formalistic) understanding of mathematics that we are not incorporating.

6 Mathematical Objectivity and the Interplay of Practices

The most important objection to a naturalistic and practice-oriented approach to mathematical knowledge comes from the perceived objectivity of mathematics. Set theorist Y. Moschovakis puts it this way:

> The main point in favor of the realistic approach to mathematics is the instinctive certainty of almost everybody who has ever tried to solve a problem that he is thinking about 'real objects,' whether they are sets, numbers, or whatever; and that these objects have intrinsic properties above and beyond the specific axioms about them on which he is basing his thinking for the moment.
> (Moschovakis [1980], quoted by Maddy [1997]: 87)

The simplest way (too simple?) of explaining this feature is by adopting realism about mathematical objects, that is to say, extending to mathematics the naïve semantics of usual natural language. That move however is rejected by naturalistic philosophers, including Maddy ([1997]), on good grounds. Thus the above quote must be contrasted with Wittgenstein's maxim: "[W]hat a mathematician is inclined to say about the objectivity and reality of mathematical facts, is not a philosophy of mathematics, but something for philosophical treatment" (Wittgenstein [1953]: 254; emphasis subtracted). The solution that naturalists are bound to explore is broadly pointed out by the slogan "objectivity without objects," formulated by Georg Kreisel and Hilary Putnam long time ago. But the real problem is in the details of the account.

According to what has been said before, in advanced mathematics we find the phenomenon of the introduction of hypotheses that are not grounded in the basic strata of mathematical knowledge, which are endowed with certainty. But, even though they are not *grounded* in them, they are *connected* with them. This interplay of practices already has consequences, since it allows us to understand (in a perfectly non-mysterious way) how it is possible for the mathematician to be *constrained* to accept certain results, even though those results concern hypothetical ingredients of mathematics. If that is the case, we can further understand how it is that different people can be said to face *the same* constraints (and that seems to explain the phenomenon of "intrinsic properties" mentioned by Moschovakis).

In the end, the notorious objectivity of mathematics would be nothing but a special and rather unique form of intersubjectivity: special because of the nature of the cognitive resources at work, very particularly at the roots of mathematical knowledge (the strata endowed with certainty, e.g., elementary arithmetic and parts of elementary geometry), but also at work in the later development of advanced practices; and special, too, because of the intimate systematic links between practices, and the way in which such links impose constraints.

In order to spell out what I mean, we shall consider a rather simple example: the introduction of concepts of infinity in set theory. The following discussion has been written more in the style of a philosopher's analysis and reconstruction, than in the style of historians, but of course the real practices that have to be explained are the historical ones. My simplified reconstruction can be regarded as an idealized model that allows us to capture crucial traits one finds exemplified in the history (not a Lakatosian "rational reconstruction").

I distinguish two quite different levels of acceptance of the infinite, those corresponding to ℕ and ℘(ℕ). One thing is to assume the existence of the infinite set ℕ of natural numbers, but it is another to take as given its power set ℘(ℕ), the set of all subsets of ℕ. The former hypothesis can be understood as committing us only to the acceptance of infinite totalities given by defining laws or rules;[23] the latter implies acceptance of infinite totalities that are arbitrary or "lawless." Thus we could speak of the contrast between accepting a tamed vs. a wild Infinity.

In a retrospective account, the First Act of set theory is the step by which we hypothetically conceptualise the never-ending collection of natural numbers as a completed set. The Second Act is the even bolder positing of power sets: In the concrete case of the natural numbers, the second hypothesis is to postulate the existence of a set comprising the totality of "all possible" subsets of ℕ. (In actual history, the first steps in the development of set theory were linked with the definition of the real numbers, and they combined rather confusedly what I've called the "First" and "Second" acts. It was only later, as a result of criticism and reconstructions, that the two different acts were made explicit and distinguished. But for the purposes of my simplified and idealized analysis, it is preferable to follow a conceptually neat order.)

The simple move of introducing a new notational convention makes it easy for us to stipulate that ℕ will now be the set of all natural numbers. We write $\mathbb{N} = \{1, 2, 3, \ldots\}$, or we employ the usual picture of a potato[24] with the symbols '1,' '2,' '3,' '...' inscribed in it. Being careful logicians, Frege and Dedekind felt rather infuriated with the trick we do by writing the dots, in an apparently clear but logically questionable move. However, a nice way to understand the meaning of '...' is simply as an *indicator* that the set-theoretic practice we are developing is meant to be systematically connected to the number practices we are very

---

[23] This could have been accepted by Weierstrass, since it complies very well with his approach to real numbers and to functions; however he had difficulty accepting bolder set-theoretic hypotheses. See Ferreirós [1999]: 34–8, 149, 183–5.

[24] So they call in France those simply connected geometric regions, enclosed by a closed curve, that we often paint on blackboards.

familiar with—and to transcend them (hypothesis) by postulating the completion of the process of formation of numbers. *Idealisation* is, thus, simply brought about by *new semiotic practices and their accompanying (explicit or implicit) axiomatic stipulations.* This makes it possible for us to introduce such things as infinite sets in what Quine liked to call our "myth-making." (There is nothing derogatory in the use of this expression: In Quine's view, myth-making is essential to all of science, and in particular to mathematics. And such a standpoint has its virtues, provided of course that one analyses carefully the *differences* between myth and math.)

Before being exposed to set theory, all of us have gained very elaborate knowledge of the natural numbers: the elementary theory of the natural numbers, and the use of natural numbers in counting. From the standpoint adopted here, I would say that we have *working knowledge of several different practices or strata, and of their systematic interconnections.* Simultaneously with the process by which we gained knowledge of these practices, we have also formed intuitive pictures in association (an example is the idea of the number line, which has been the topic of much research recently).

The fact that the new concepts and hypothetical assumptions link back with previous knowledge has consequences. If we accept the hypothesis that there is a set of all natural numbers, it is almost inevitable to admit as well the existence of any *definable set* of natural numbers. These are also infinite (or finite) totalities given by well-determined rules or defining conditions, obvious examples being:

– the set $\mathbb{E}$ of even numbers;
– the set of multiples of 391; and
– the set of prime numbers.

Notice that not all "possible" infinite subsets of $\mathbb{N}$ are definable subsets. Since the latter are proper parts of $\mathbb{N}$, which by assumption is a set, that can be defined explicitly by means of antecedent properties and relations (i.e. such that they do not depend at all on the new hypotheses), a decision not to acknowledge them would be close to contradictory. Whatever is predicated of all of $\mathbb{N}$, has to be predicated of such definable subcollections; if $\mathbb{N}$ is regarded as a completed domain, so must be

E, etc.; if N is regarded as a class-as-one, a set, so must be its definable subsets.

But now, our *antecedent* arithmetical knowledge has the implication that there exist one-to-one correspondences between N and each of the subsets just mentioned. Previous mathematical knowledge entails that such mappings exist, for, corresponding to the three examples above:

- The operation $y = 2 \cdot x$ now determines a bijection, $f: \mathbb{N} \to \mathbb{E}$.
- Similarly for $y = n \cdot 391$.
- Euclid's proof that no finite collection of primes contains all the prime numbers, has the consequence that for each n there is an nth prime number.

This is precisely one of the ways in which math is different from myth. The constraints to which mathematical "myth-making" is subject are undeniably much, much tighter than those valid for mythology. Once we accept the mathematical hypothesis of admitting N (and its definable subsets), we are *forced* to admit that there are bijections between N and each of its (definable) infinite subsets.

This, by the way, affords an explanation of the possibility of *simultaneous discovery* in mathematics: Both Dedekind and Cantor discovered that simple fact, independently as it seems, during the 1870s. Galilei could not have discovered the "fact" because he was not accepting the same hypotheses as Dedekind and Cantor did 250 years later. On the assumption that infinities have to be conceptualised as magnitudes, the above arguments present us with a contradiction, as Galilei realized: since he accepted Euclid's principle that a whole magnitude is greater than its parts, they became the basis for the conclusion that there cannot be actual infinities (as magnitudes). On the basis of the very different assumption that infinities must be conceptualised as sets, those same arguments force us to admit that part and whole can be bijected, and that the set is greater than the subset in the sense of proper inclusion, but not greater in the sense of bijection.

This situation is merely *discovered* on the basis of previous knowledge and the new hypothesis. N may be called an "invention," but it leads to "discoveries." There is nothing paradoxical in this, which is merely a consequence of the interplay between strata of knowledge, or practices. The traditional dichotomy invention/discovery is simply inadequate to deal with the conceptual subtleties of the mathematician's handling of hypotheses. To analyse what is going on in the development of mathematical theories by the introduction of new hypotheses, one would need at the very least a trichotomy. Pure invention is almost never present in mathematical developments, but invention *cum* discovery is typically present whenever a new hypothesis is introduced.

The second hypothesis, what I have called the Second Act (the Power Set Axiom as restricted to $\wp(\mathbb{N})$), is already of the highest importance: It leads us into Cantor's continuum problem. But once again, a deep reflection on its connections with previous mathematical knowledge and practices makes clear a number of implications and results. To see this, we would have to fill in our general scheme with some rich details about mathematical practices as they existed in the early 19th century. E.g., it was characteristic of the situation at that point that the non-negative reals were regarded on an equal footing with the natural numbers; this was sanctioned by a long tradition going back to the identification between number and proportion in the early 17th century. For this reason the tendency was that, whatever was predicated of $\mathbb{N}$, was also applied to the domain of the positive reals.

The links between the new concepts and assumptions and previous knowledge again had consequences: Well-established knowledge about the real numbers led to the result (again a "discovery") that $\mathbb{R}$ and $\mathbb{N}$ are not bijectable, $\mathbb{R}$ is not denumerable. Once we have adopted the idea that $\mathbb{R}$ is also an infinite set, *it is not arbitrary, but necessary* (in the light of previous mathematical knowledge), to conclude that $\mathbb{R}$ has a greater cardinality than the set $\mathbb{N}$. In particular, the details of Cantor's reasoning in [1874] employ only mathematical results that were available before 1850, even though they are employed in the pursuit of questions that were posed only later.

Furthermore, the bold hypothesis associated with the Second Act of set theory (wild Infinity) was not quite arbitrary either: When we consider the historical context of established mathematical practices at the time, we realize that the traditional assumptions about the real numbers (and also Dirichlet's concept of function) constituted clear arguments for the adoption of this second hypothesis, if the first was to be accepted at all.[25]

7 Conclusion

In the last section I have tried to make clear the way in which a naturalistic, practice-oriented approach to mathematical knowledge, such as the one I am proposing, can explain the objectivity of mathematical results. To do so, it has been necessary to discuss an example in which we find hypothetical elements involved, since those are the cases that really pose a problem. As we have seen, it is through the interplay of different coexisting practices that constraints on the development of new strata of mathematical knowledge enter the scene. In our examples, we have seen systematic interconnections between assumptions about infinite sets and pre-existing practices that have to do with the natural and the real numbers. As the example of Cantor has made clear, my claim that those links guide the development of mathematics is not a philosophical assumption that we superimpose on the actual developments, but rather it makes explicit an integral element of actual practices of conceptual and theoretical development.

Obviously, the foregoing discussion has made no attempt to analyze completely the web of practices involved in our examples. In order to fully elaborate the last example, we would have had to describe in detail the mathematical practices having to do with the concept of the real numbers, as they existed in the early 19th century; their links with assumptions about magnitudes and with the theory of proportions; and the links of the latter with practical problems of measurement. Hopefully the foregoing has been sufficient to enable the reader to make for herself a picture of what this full account might look like (a scheme of several different strata of mathematical knowledge, ranging from technical practices to advanced mathematics, systematically interrelated with

---

[25]Lack of space prevents me from going into the details, which will be given in a book that I am currently writing; it is provisionally named *Mathematical Knowledge and the Interplay of Practices*.

each other), and I hope that it will thus may have allowed the reader to begin to appreciate the explanatory power of the present approach.

When we consider the whole range of practices that coexist and interrelate, it becomes possible to make an analysis of the cognitive and cultural elements that join each other to form the intricate puzzle of mathematical knowledge. As should be clear from the previous discussion, my understanding of the "cognitive" is more in agreement with current tendencies in psychology, neuroscience, and biology, than with traditional perspectives in AI-style cognitive science. It should also be clear that I am inclined to assign an important role to language, and to the written symbols, as cognitive elements making possible the elaboration of mathematical knowledge. And I have made explicit that, in my opinion, technical practices have a particularly important role to play, at the very roots of mathematical knowledge. The resulting picture is a complex one, in which many different elements interweave to form a peculiarly solid fabric. Each and every one of the elements and steps is asking for more detailed analysis.

Bibliography

Bottazzini, U. [1986]. *The* Higher Calculus*: A History of Real and Complex Analysis from Euler to Weierstrass.* Berlin and New York: Springer.

Campbell, J.I.D., ed. [2004]. *Handbook of Mathematical Cognition.* Psychology Press.

Cantor, G. [1874]. "Über eine Eigenschaft des Inbegriffes aller reellen algebraischen Zahlen." *Mathematik* **77**. 258–62.

Dedekind, R. [1888]. *Was sind und was sollen die Zahlen?* Braunschweig, Vieweg.

Dehaene, S. [1997]. *The Number Sense: How the mind creates mathematics.* Oxford University Press.

Dolezal, H. [1982]. *Living in a world transformed.* Chicago: Academic Press.

Ferreirós, J. [1999]. *Labyrinth of Thought: A history of set theory and its role in modern mathematics.* Paperback edition, 2007, with a new Epilogue. Boston: Birkhäuser and Basel.

— [2005a]. "Certezas e hipótesis: perspectivas históricas y naturalistas sobre el conocimiento matemático." *Filosofía de las ciencias naturales, sociales y matemáticas.* Ed. by A. Estany. Madrid: Trotta.

— [2005b]. "Dogmas and the Changing Images of Foundations." *Philosophia Scientiae.* Ed. by G. Heinzmann and P. Nabonnand. Cahier spécial Fonder autrement les mathématiques 5. 27–42.

Ferreirós, J. and J. Gray, eds. [2006]. *The Architecture of Modern Mathematics: Essays in history and philosophy.* Oxford University Press.

Hurford, J. [1987]. *Language and Number: the emergence of a cognitive system.* Oxford: Basil Blackwell.

Kitcher, P. [1984]. *The Nature of Mathematical Knowledge.* Oxford University Press.

— [1988]. "Mathematical Naturalism." *History and Philosophy of Modern Mathematics.* Ed. by W. Aspray and P. Kitcher. University of Minnesota Press. 193–325.

Kohler, I. [1964]. *The formation and transformation of the perceptual world.* New York: International University Press.

Lakatos, I. [1967]. "A rennaissance of empiricism in recent philosophy of mathematics?" *Mathematics, science and epistemology.* Vol. 2. Cambridge University Press. 24–42.

Maddy, P. [1990]. *Realism in Mathematics.* Oxford University Press.

— [1997]. *Naturalism in Mathematics.* Oxford University Press.

— [2007]. *Second Philosophy: A Naturalistic Method.* Oxford University Press.

Manders, K. [2008]. "Diagram-based Geometric Practice." *The Philosophy of Mathematical Practice.* Ed. by P. Mancosu. Oxford University Press.

Moschovakis, Y. [1980]. *Descriptive set theory.* Amsterdam: North-Holland.

Piaget, J. [1970]. *L'épistémologie génétique.* Paris: Presses Universitaires de France.

Pica, P. and S. Dehaene [2006]. "Core Knowledge of Geometry in an Amazonian Indigene Group." *Science* **311**. 381–4.

Pickering, A. [1995]. *The Mangle of Practice: Time, agency & science.* University of Chicago Press.

Quine, W.V.O. [1969]. "Epistemology naturalized." *Ontological Relativity and other essays.* New York: Columbia University Press.

Riemann, B. [1868]. "Über die Hypothesen, welche der Geometrie zu grunde liegen." *Gesammelte Werke und wissenschaftlicher Nachlass.* Republished in 1990. Berlin: Springer. 272–87.

Tappenden, J. [forthcoming]. *Philosophy and the Origins of Contemporary Mathematics: Frege and his Mathematical Context.* Oxford University Press.

Weyl, H. [1927]. "Diskussionsbemerkungen zu dem zweiten Hilbertschen Vortrag über die Grundlagen der Mathematik." *From Frege to Gödel.* Ed. by J. van Heijenoort. English translation published in 1967. Harvard University Press. 480–84.

Wilder, R.L. [1981]. *Mathematics as a Cultural System.* Oxford: Pergamon Press.

Wittgenstein, L. [1953]. *Philosophical Investigations.* Oxford: Basil Blackwell.

ESSAY 7

Pragmaticism as an Anti-Foundationalist Philosophy of Mathematics

Ahti-Veikko Pietarinen

According to Charles Peirce's pragmaticism, mathematical reasoning concerns the creations of the mind. It emphasises experimentation and observation on diagrammatic and iconic representations of these creations, and does not presuppose mathematical foundations. This paper contrasts Peirce's pragmaticism with a number of recent philosophies of mathematics, including intuitionism, structuralism, fictionalism, platonism, and quasi-empiricism. I argue that pragmaticism, as an anti-foundationalist philosophy of mathematics, is a positive thesis about such diagrammatic and iconic representations drawing from actual mathematical practice.

Peirce's remarks on the philosophy of mathematics are not well known, and appear among his actual mathematical and logical works. Many of them have not been made public to date. Some of these remarks Peirce never explicated with any particular regard for the philosophical aspects of mathematics. Rather, they relate to what he sees essential in actual mathematical practices, and as such are not separate from his extensive research on logic, semeiotic, phenomenology, and the methodology of special sciences. After all, the entire industry christened as the 'philosophy of mathematics' has been post-Peircean.

This ought not to block our philosophical road to Peirce's inquisitive thoughts in the least, however. In this paper I wish to establish that what Peirce has to say about the philosophy of mathematics is very significant. My task is to study what he had in the offing in comparison with what we nowadays regard as pivotal research questions falling within the philosophy of mathematics. Such a study enables us to put his views into a sharper focus than what might be possible by merely trying to understand his views in their own right, or as if his handiwork were only of some historical or exegetic interest. As a by-product I hope to communicate how the comparison enables us to appreciate when something goes wrong in the received theories concerning the philosophy of mathematics and mathematical practice.

1 Pragmaticism as an Anti-Foundationalist Philosophy of Mathematics

I have argued in Pietarinen [forthcoming] that Peirce's pragmaticism articulates a self-standing philosophy of mathematics which differs markedly from the foundationalist philosophies such as logicism, intuitionism, and platonism. It differs from Hilbert's axiomatic programme and quasi-empiricism in notable respects, too. A term of art that might be used to characterise Peirce's position is *anti-foundationalism*. As such it does not explain much, however. In which sense is pragmaticism not a foundationalist philosophy of mathematics? What is Peirce's philosophy of mathematics? What is its positive contribution? These questions define my agenda in the present treatise.

A couple of remarks that Peirce makes on the nature of mathematics are particularly relevant to our concerns. In 1896 he noted the following:

> It is an error to make mathematics consist exclusively in the tracing out of necessary consequences. For the framing of the hypothesis of the two-way spread of imaginary quantity, and the hypothesis of Riemann surfaces, were certainly mathematical achievements. Mathematics is, therefore, the study of the substance of hypotheses, or mental creations, with a view to the drawing of necessary conclusions.
> (NEM 4:268, 1896, *On Quantity, with Special Reference to Collectional and Mathematical Infinity*)

The substance of mathematics refers to mental creations. Despite this allusion, which might, sight unseen, suggest that Peirce's thought is allied with intuitionistic philosophy, I shall argue that there are compelling reasons to take Peirce's views to be quite remote from the concerns of intuitionism. After all, intuitionism is a foundationalist philosophy which takes mathematical objects to be created or constructed by our mental and cognitive processes. It is the aspects of mathematical reasoning, Peirce explains a couple of years later, that are hypothetical and whose constructions refer to the "creations of the mind":

> Mathematical reasoning holds. Why should it not? It relates only to the creations of the mind, concerning which there is no obstacle to our learning whatever is true of them . . . It is fallible, as everything human is fallible. Twice two may perhaps not be four. But there is no more satisfactory way of assuring ourselves of anything than the mathematical way of assuring ourselves of mathematical theorems. No aid from the science of logic is called for in that field.
> (CP 2.192, c.1902, *General and Historical Survey of Logic: Why Study Logic?*)

Mathematics and logic are strictly separate sciences, as Peirce always insists, having learned that notion from his father, Benjamin Peirce. Mathematicians *reason*, while logicians are engaged in the *study* of the processes and theories of reasoning.

That proper mathematical reasoning "relates only to the creations of the mind" does not mean, unlike in intuitionism, that it is mathematical objects that are created in the minds of those who practice mathematics. It means that in studying the outcomes of the mental processes correlated with mathematical reasoning we are at once engaged in studying those aspects of the mathematical structures in which real mathematical objects are seen to figure. Mathematical signs have objects and interpretants just as any other kinds of signs do. Unapproachable by the intuitionist philosophy of mathematics, Peirce succeeds in avoiding unwarranted appeals to psychologism and mentalism.

Another observation cropping up in the previous quotation is the fallible character of mathematics. I will comment on fallibilism in mathematics in a moment; what is clear by now is that Peirce's disdain for foundationalist approaches to mathematics does not imply rejecting or denying anything of the relevance and reality of mathematical *facts* just as it does not mean rejecting or denying the reality or objectivity of mathematical *objects*. The denial of anything of this kind would mean subscribing to a substantial foundational position, such as nominalism, in order to be able to explain at all of what such a negation is in fact taken to consist. Accordingly, one conclusion I argue for in this paper is that pragmatistic stance disavows those off-shoots of structuralist philosophies of mathematics that see mathematical objects as objects of fiction (Field [1980]).

Pragmatistic philosophy recognises that (i) laws of logic, just as laws of nature, are subject to change, and that (ii) no single science is capable of founding all others. In mathematics, logic serves no foundational purpose.[1] In fact, in Peirce's classification of the sciences, mathematics and phenomenology are sciences that precede normative logic (Pietarinen [2006a]), and mathematics is not grounded on anything else. In a very concrete sense, thus, it is mathematics that provides the *a priori* for metaphysical investigations.

Pragmatistic philosophy of mathematics accentuates the importance

---

[1] Blais ([1989]) has characterised a kind of anti-foundationalist pragmatic philosophy of mathematics that might have satisfied William James but probably not Peirce, since it overlooks the finesse of Peirce's position. Patin [1957] is one of the earliest studies on Peirce's philosophy of mathematics. Its representation of the main tenets of pragmatism, intuitionism, and formalism is entirely obsolete, however.

of the actual practice of mathematics. At the same time, it displays an astute recognition of the existing aperture between the actual practices of mathematicians on the one hand and mathematical systems (axiomatic, formal systems) on the other. Since Charles, just as Benjamin Peirce, was an accomplished mathematician, he held a privileged vantage point from which to observe the existence of such a gap and the reasons for its existence. One of his observations concerns the interpretation of Aristotelian axiomatic method: that such a method is bound to be quite foreign to mathematical practices if it is taken to constitute the bulk of what mathematicians do, namely to deduce (by syllogistic reasoning or otherwise) theorems from a given collection of axioms.

Being a working mathematician, it might also appear puzzling why Peirce was not a platonist regarding the nature of mathematical knowledge and mathematical objects. Briefly, the reason is methodological: Mathematics suggests deep metaphysical questions that need serious philosophical analysis, while the quest for answers must be guided by an insight into the nature of mathematics as well as a broad application of concrete mathematical practices. I will come back to platonism in a separate section below.

Nor can Peirce's approach to the metaphysics of mathematics reasonably be claimed to be naturalistic, either. The nature of mathematical entities is wholly different from the nature of the entities, including laws or principles governing mathematical discoveries, that are established by the methods of natural sciences.[2]

What Peirce would not say, it is appropriate to emphasise, is that these mathematical practices are necessarily social. Popular treatises tend to emphasise that in order for any true scientific inquiry to make progress, it needs to go through the common and shared public investigative efforts of the entire scientific community. But mathematical

---

[2]Kitcher's ([1984]) naturalistic and quasi-empiricistic approach to mathematical objects is a case in point of unwarranted naturalisation of mathematics from the Peircean practice-based point of view. Hoffman ([2004]) suggests that Kitcher's approach be 'fictionalised' by turning the notions Kitcher uses as examples of idealisation in mathematics and science into characters of fiction. This proposal contains a fallacy of moving from the non-existence of idealised objects to fictional objects, however. Why it is a fallacy is shown, among others, by its unviable consequence: It would establish the fictitious character of the whole of the objects of science in the same go. For a critique of fictionalism, see a separate subsection below.

facts cannot be results of inherently social practices. What matters is that mathematicians act in accordance with the habits of reasoning in a certain way in response to certain kinds of mathematical problems. The cultivation and modification of these habits does not rest on social factors but on self-controlled action guided by the ideals that mathematicians contemplate in their minds in dealing with mathematical problems.[3] Peirce calls such an instinctive and stable faculty of reasoning the inquirer's *logica utens*. It is with such a *utens* that mathematicians are able to prove theorems, not by the *logica docens*, which is the schooled and developing faculty of theories of reasoning (Pietarinen [2005]). Peirce explains this matter in the passage immediately preceding the quotation above concerning the nature of mathematical reasoning:

> You think that your *logica utens* is more or less unsatisfactory. But you do not doubt that there is *some* truth in it. Nor do I; nor does any man. Why cannot men see that what we do not doubt, we do not doubt; so that it is false pretence to pretend to call it in question? There are certain parts of your *logica utens* which nobody really doubts . . . The truth of it is too evident. Mathematical reasoning holds. Why should it not? It relates only to the creations of the mind. (CP 2.192)

A characteristic feature of pragmaticism is that there is no and need not be any ultimate basis for knowledge. It is the basic property of Peirce's notion of *continuity*, which characterises the "true continuum," that all cognition rests on former cognitions but that there is not and need not be any first cognition or first transcendental object. Yet his intuitive concept of such a continuum has eluded mathematical definitions. We cannot even begin to survey the attempts that have been made to that effect here,[4] but let us note that Peirce thought true continuum to be created by a peculiar phenomenological insight rather than by any

---

[3] See here Peirce's unpublished MS 280, 1905, *The Basis of Pragmaticism*. Transcription available at http://www.helsinki.fi/~pietarin/.

[4] See e.g. Hudry [2004], Myrwold [1995], Ehrlich [forthcoming], and Stjernfelt [2007] for recent attempts and expositions.

well-definable mathematical structure. This does not mean that an approximate mathematical model could not be found for it. Such a model may well be within the reach of contemporary metamathematics, in so far as it is taken into account that the investigation must take place in the borderlands of mathematics (the 'first' science of discovery for Peirce according to this classification of the sciences) and phenomenology (the 'first' philosophy and the 'second' science of discovery).

This understanding of continuum objectifies *fallibilism*, the view that our current theories of science, including our mathematical theories concerning mathematical facts, may turn out to be false. Reasoning and observation in science is performed by us, human beings. Products of science follow from the methods employed in reasoning and observation. Science, including exact sciences, are anthropomorphic in the sense that we never acquire absolute certainty concerning the truth of our best scientific theories. Since Peirce's continuum adds to actuality all that is possible—including all possible mathematical entities, all possible objects, relations, propositions, and facts—and since possibility greatly outweighs actuality, there will be an inevitable uncertainty and vagueness in reality that cannot be disposed of even by the best theories and the best methods of exact sciences currently at hand. The point is recorded here:

> The principle of continuity is the idea of fallibilism objectified. For fallibilism is the doctrine that our knowledge is never absolute but always swims, as it were, in a continuum of uncertainty and of indeterminacy. Now the doctrine of continuity is that *all things* so swim in continua.
> (CP 1.171, c.1897, *Notes on Scientific Philosophy*)

It is from this *synechistic* and modal nature of the true continuum that fallibilism of our best theories ensues, not vice versa. However, since the model of synechistic mathematics is something that is not and need not be 'well founded', the fallibilistic epistemology concerning mathematical truths receives its explanation without any need of seeking being founded on anything else.[5]

---

[5]One might contemplate models of non-well-founded set theory as the suitable candidates for Peircean continuum. Even if there be some analogy, as such non-well-

2 Which Philosophy of Mathematics Is Pragmaticism Not?

Given these introductory remarks on Peirce on the nature of mathematics, I will next contrast his views with some of the subsequent work on the nature and practice of mathematics.

2.1 INTUITIONISM

A couple of years after Peirce, Hermann Weyl ([1987]: 119), and L.E.J. Brouwer ([1975]) arrived at a position that seems similar to the Peircean continuum: The continuum is "intuitive" and to be conceived "as-a-whole." Brouwer's view differs markedly from Peirce's in crucial respects, however. According to Brouwer, "Mathematics is a free creation, independent of experience; it develops from one single a priori Primordial Intuition" (*On the Foundations of Mathematics* 1907: 179, in Brouwer [1975]). This contrasts sharply with Peirce in three respects. First, mathematics is not independent of experience since experience and theory are not separable in pragmaticism.[6] Second, any appeal to Cartesian intuition or a Kantian transcendental object as the final resort by which mathematical knowledge is fashioned is rejected by a pragmaticist.[7] Third, the mind is not for Peirce an individual or single creating or constructing solipsist "Subject" capable of intuition, but a collective and general "creatory" (MS 318: 18) of all kinds of signs, including signs standing for mathematical ideas. Consequently, mathematics is not absolutely "free creation"; it proceeds according to the habits of thinking

---

foundedness it is not enough. It suffers from the dubious notion of a set which Peirce wanted to replace with the notion of a collection, which is a "derivative individual" and does not have any "characters" attached to it (*Peirce to Cantor*: 2, 21 December 1900; Pietarinen [forthcoming]). Moreover, non-well-founded sets give rise to point-like structures just like ordinary axiomatic set theories with the axiom of foundation do and thus cannot agree with the true continuum.

[6] Experience of course does not mean only sense experience.

[7] The sole idea of Descartes that both Peirce and Brouwer might have agreed with is that by logic we do not create new mathematical truths, since according to both, logic is the science which studies the processes of drawing necessary conclusions, whereas mathematics differs from logic in being the science which draws necessary conclusions.

and reasoning, which are real generals linking contexts of mathematical discovery with action.

Yet it is true that Brouwer's concept of "the Intuitive Continuum" has some resonant similarities with Peirce's. For instance, in extending the continuum with the possible alongside the actual Brouwer broadened his philosophical horizon and did not remain an actualist about mathematical objects. But for Brouwer, the possible stood for "identifiable points," whereas Peirce rejected the notion that the "*would-bes*"—those preliminary conceptions of what might or could happen when we enquire about the world, or the inner thought, or the parts of the worlds or thoughts connected with our mutually agreed universes of discourse—can possess any point-like identities.

Consequently, Murphey ([1961]) and Engel-Tiercelin ([1993]) are mistaken in holding Peirce's views to stand in strict opposition to intuitionism because of intuitionism's commitment to actuality only. In Brouwer's formulation, possibilities are admissible constituents of the continuum. However, their metaphysical status and nature was quite different from Peirce's realistic conception influenced by his reading of the scholastic philosophers. According to Peirce's scholastic realism, what is possible is as real as what is actual. Unlike Brouwer, Peirce did not wish to extend the continuum with the possibilia merely to resolve set-theoretic paradoxes.[8]

Nevertheless, some similarities do take place between intuitionism and pragmaticism (Pietarinen [2006b] and [forthcoming]). Both take mathematical reasoning as related to the creations of the mind. The law of excluded middle is rejected as an a priori logical principle.[9] Existence is subordinate to inferential and cognitive processes and to the activities of seeking and finding. Unlike intuitionism, however, Peirce did not consider the failure of the law of excluded middle to be a failure of

---

[8] Peirce's remarks on set-theoretic paradoxes come from an incomplete note "Mr. Bertrand Russell's Paradox" (MS 818, 1911, 5 pages). In that note, he takes the sentences giving rise to an alleged paradox "to be easily interpreted so as to remove their contradictory sound" (ibid.: 3). Whether he had a predicative re-interpretation in mind of what the sentences define is impossible to tell, since the note ends abruptly before we get a comprehensive exposition of the suggested solution, and the remaining pages have been lost.

[9] Peirce studies indeterminate properties in terms of the limit interpretation.

proving or being able to provide a *construction* for one of the disjuncts, but that there are situations (models) in which none of the propositions can be *asserted*.

2.2 STRUCTURALISM

Peirce's position on continuity makes his philosophy of mathematics incompatible also with mathematical structuralism. According to (*ante rem*) structuralism (Shapiro [1997]), objects of mathematics are identified by their positions or roles that they have in various kinds of mathematical structures. Hence the subject of study in the philosophy of mathematics is the general account of all kinds of structures, such as the study of classes of models or theories of manifolds.[10]

Structuralism displays some likeness to pragmaticism. What Peirce takes to be real in mathematical constructs, in the sense of real being that which is independent of what anybody thinks of it, are the relations in hypothetical creations of the mind. Likewise, structuralism, especially category-theoretic structuralism of Hellman ([2004]), takes mappings or morphisms to be the first-class entities of mathematics, and mathematical objects to emerge only derivatively. Peirce took forms of relations to be all-important in constituting the subject matter for the logical investigation of the category of secondness, namely that of what exists and what is actual. This derivative nature of objects with respect to mappings means that their identities follow from the locations or places those objects have in the "mental chart"[11] created by mathematical imagination. In this sense, therefore, to take the identities of objects to draw from the locations or places those objects have on a map, chart, or structure is to subscribe to a version of structuralism concerning the primacy of relations over objects in mathematical ontology.

In the end, however, pragmaticism turns out to be incompatible with structuralism. The latter carries with it a nominalistic assumption of

---

[10] Since time is a continuum in which one can count even though there is no natural notion of a successor in time, mathematical structuralism should not be confounded with Kantian structuralism, a nominalistic philosophy of mathematics that Benacerraf ([1973]) has defended.

[11] MS 280, 1905, *The Basis of Pragmaticism*.

atomicity of mathematical objects. Given Peirce's characterisation of the true continuum, however, individual identities dissolve in it because it "is something whose possibilities of determination no multitude of individuals can exhaust" (CP 6.170, 1901, *Synechism*). There are neither individual points nor elements constitutive of continuity, and hence positions for those points or elements cannot be identified in such continuum. And so from the synechistic (and inter alia pragmatistic) point of view, the basic doctrine of structuralism is not satisfied. Therefore, from the overall pragmatistic point of view structuralism is bound to fail as a foundational enterprise for mathematics.

There are other versions of structuralism, in particular the modal structuralism of Hellman ([1989]), that come closer to pragmatistic concerns. Modal structuralism strives to dispense with those nominalistic assumptions of structuralism that appeal to atomistic postulates. However, it is still a version of category-theoretic structuralism, and it is likely that, because of its reliance on collections of mappings and structures as point-like constructions, Peirce would have regarded it as a form of nominalism. The argument certainly deserves longer exposition, as the conclusion hinges on the notion of morphism as an explication of general properties of mathematical domains as well as the concept of continuity involved in category theory. Nonetheless, in an undisputable sense category theory is just like set theory as the distinction is routinely drawn between small and large structures.

Hellman ([2004]) suggests overcoming these limitations and set-theoretic commitments by reformulating mathematics topos-theoretically, providing its own universe of discourse, which arguably is no longer a set-theoretic one. Without discussing here the point as to whether this logical approach to category theory can really do without set theory, we note that, interestingly, topos theory appears to provide a mathematical theory of *iconicity* in terms of homomorphism between domains and co-domains analogous to, for instance, continuous maps between topological spaces. Notable here is that the line of research that Hellmann has been pursuing is a foundationalist approach to mathematics and not so much an alternative philosophy to it.

Modal structuralism, which attempts to dispense with the remaining nominalistic undercurrents of ordinary category-theoretic structuralism,[12] nevertheless coheres with Peirce's pragmaticism in certain respects. Hellman seeks to avoid the set-theoretic commitment to the totality of the universe of mathematical objects. Mathematical domains are indefinitely extendible and relative to possible worlds in the sense that mathematical constructions talk about hypothetical constructions. Mathematics, according to Hellman ([2004]: 146), concerns "what would necessarily be the case were the relevant structural conditions fulfilled." Peirce took mathematical reasoning to be of this subjunctive form while concerning necessary reasoning under such hypothetical constructions. Hellman (ibid.) takes mathematics to make "no actual commitment to objects at all, only to (propositionally) what might be the case." Like Peirce, he takes mathematics to concern possible objects. He then suggests an axiomatisation of modal existence by axioms of second-order modal logic. Like Peirce, Hellman takes actual mathematical practice to guide the axiomatisation of theories. Here it means looking at the mathematical practice that could determine what the axioms for the S5 system of second-order modal logic might look like. A question not to be forgotten is that, if those axioms and the language of S5 are couched in set-theoretic terms, does it imply that their origin is in the axiomatic set theory? Hellmann says little about the semantics for S5 for apparent reasons.

From having such a very strong logic at his disposal, Hellman encounters similar issues as Peirce did in attempting to make sense of the identities between what is actual and what is possible.[13] Hellman thinks that we cannot in fact have quantification over relations or similar higher-order entities since that would commit us to identifying actual relations with possible relations. Hence, he does not accept cross-identification with respect to higher-order notions and takes each possible world to constitute its own mathematics.

Hellman's theory opens up significant perspectives on the indispensability of higher-order notions such as relations, functions, and map-

---

[12] Hellman's proposal continues what Bell ([1988]) suggested in replacing the absolute universe of set theory with the plurality of the universes of topoi, each providing a possible world in which mathematics is made.

[13] See Peirce's 1906 *Prolegomena to an Apology for Pragmaticism*; CP 4.530–72; and Pietarinen [2008].

pings as the building blocks of fundamental mathematical ontology.[14] However, we might ask whether his proposal runs into serious problems when the *epistemology* of mathematics entities is at issue. Without cross-identification, we are denying that these relations, functions, or mappings are genuine many-world entities. From the metalogical perspective, we would be prevented from knowing what or which relations or functions they in fact are. Such a denial means depriving mathematical knowledge of some of its key subject matters. And there are credible grounds to take the indispensability of such higher-order notions to also mean our mathematical knowledge of them. Mathematical knowledge requires their identification. Hence, what lurks around the corner after all in Hellman's account is nominalism with respect to relations and functions.

Hellman's proposal differs from Peirce in that Hellman does not take possible objects to be real possibilities in the sense of being the real constituents of the objective world that includes mathematical entities and facts. This point follows from his rejection of the possibility of cross-identification of higher-order notions. Hellman indeed admits that his talk of possible worlds is "heuristic only" and that there are "literally no such things" such as those that merely might have existed (Hellman [2004]: 147). His second-order comprehension schema applies only "within a world" (ibid.). In contrast, Peirce had no such scruples and took real possibilities to be true constituents of the world. His scholastic realism, which lies at the core of pragmaticism, applies to mathematics just as to metaphysics. Pragmaticism, Peirce states, "is most concerned to insist" upon "the reality of some possibilities." (CP 5.453, 1905, *Issues of Pragmaticism*), which is precisely what Hellman denies.

Consequently, pragmaticism cannot be seen compatible with modal structuralism, either. However, developing a realistic alternative to Hellman's modal-structuralist mathematics, with the notion of an identification of second-order entities across possible worlds guided by mathematical practice, might turn out to be one of the closest contemporary counterparts to the pragmatistic restatement of the philosophy and practice of mathematics.

---

[14]Colyvan ([2001]) discusses the indispensability arguments in mathematics and science.

2.3 FICTIONALISM

Burgess ([2004]: 19) claims that modal accounts of mathematics pertain to another off-shoot of structuralism known as fictionalism. From the pragmatistic perspective this is an oversimplifying assimilation. Peirce's scholastic realism takes possibilities to be just as real as actualities and having nothing to do with fiction. What is real, including a real possibility, is in fact a contrary to what is fiction: According to Peirce, "[I]f it be not real it can only be fiction: a Proposition is either True or False" (CP 4.547 and *Prolegomena*).

Yet in the literature a mini-industry has emerged holding that on the ruins of structuralism another nominalistic philosophy can be erected (Field [1980] and [1989]). The bottom line is that mathematical statements are not about just any mathematical thing at all. Mathematical statements build up narratives, and just as narrative stories of fiction, they give rise to fictional entities. Since there are no objects, mathematical statements are strictly speaking all false.

From the pragmatistic vantage point, such an extreme nominalism means blocking the road to mathematical inquiry. How can fictionalism account for progress in mathematics?[15] What is the ontology of fictional entities?[16] Peirce would have not accepted a fictionalist way of looking at mathematics. His remarks on fiction and the use and meaning of fictional names are plentiful, but they never bear on mathematics: "The fictive is," he tells, "that whose characters depend upon what characters somebody attributes to it; and the story is, of course, the mere creation of the poet's thought" (CP 5.153, 1903, *Three Types of Reasoning*). But mathematics is not a story created by somebody, namely a mathematician, just as the identity of mathematical objects cannot depend solely on what characters this somebody happens to attribute

---

[15]Recall here the 'no miracle' arguments in the philosophy of science. Consistency is not a sufficient requirement, since there are uncountably many 'stories' consistent with any previous installment of the 'story.' On the other hand, inconsistency is not in such bad books as philosophers of mathematics might have thought. Often, discovery and progress in mathematics consists precisely in the capability of hitting on right kinds of inconsistencies.

[16]Ontological maxims to save fictionalism, such as requirements of economy or parsimony of fictional entities do not help, because mathematics does not ascribe to such maxims, mathematics simply is not simple.

to them. The identities of mathematical objects must be free from such singular attributions.

Moreover, if the ontology of mathematics corresponds to fiction as novels or narratives correspond to fictional objects, as it according to fictionalism does, mathematics would fall short of serving as the bedrock science of discovery from which other fields of sciences of discovery, namely philosophy and the special sciences, draw their inquisitive inspiration (I am referring here to Peirce's perennial classification of sciences, see Pietarinen [2006a]). Given the order of dependence as the one Peirce sketched in his classification of the sciences, fictionalism forces the ontological status of the objects of all sciences to be fictional. However, according to fallibilism that characterises pragmaticism, not all scientific statements can be false in one go, although any one of them may turn out to be subject to revision or rejection in the long run.[17]

There are other arguments against fictionalism that can be marshalled. One begins with the Brouwerian premise that mathematics is not a language at all, that mathematics has nothing to do with language. Since fictionalism presupposes the possibility of narratives, and since narratives are necessarily linguistic, from the point of view of Brouwer's intuitionistic, languageless mathematics, fictionalism is unattainable.

In conclusion, fictionalism is not a plausible alternative to structuralism. It mistakes the view that mathematical reasoning concerns imaginary or phenomenal objects created by the processes of the mind for the view that takes these imaginary or phenomenal objects to be on a par with fiction.[18]

2.4 PLATONISM

Some take fictionalism, and incorrectly to my judgment, to be a contemporary recasting of platonism. My question is phrased as one concerning the relationship between platonism and pragmaticism. Since Peirce was a working mathematician and a scholastic realist about mathematical

---

[17] Taking refuge in conservatism about mathematics does not help here, either, since pragmaticism does not accept sciences of discovery to rest on a nominalistic basis.

[18] Thomas ([2007]) presents some more cautions reservations concerning the coherence of fictionalism in mathematics.

entities, it might appear that he would have subscribed to some version of platonism. This was not the case, however. According to that heterogeneous idea differently characterised over decades of dispute among philosophers of mathematics, what mathematics deals with are abstract objects that exist independently of our ways of conceiving them. Existence takes place in the totality of the universe within which our actual world obtains an accidental location. Abstract objects and mathematical forms linger in that external world and are up for grabs to be picked out or defined by the hardworking mathematician who is on the lookout for them. Yet they are something neither material nor mental. And if so, how can we know of such things?

According to pragmaticism, mathematical objects are, indeed, very concrete and real, tangible objects, but at the same time the manifold ways of conceiving them are amenable to constant experimentation and observation. The objects do not exist as platonic forms are taken to exist. Actuality and existence are processes undergoing continuous maintenance. They are "occurrences" in the universes of discourse.[19] Existence comes in degrees, and there are countless kinds of existences. For one, existence requires identification. Therefore Peirce, despite his deep respect for many aspects of Plato's philosophy, could not have taken seriously mathematicians' routinely expressed belief—almost a blind faith—in a pre-existing realm of mathematical abstractions that is somehow ready-made and waiting to be discovered.

That a mathematician's attitude is typically platonistic was aptly recognised by Peirce early on:

> If you enjoy the good fortune of talking with a number of mathematicians of a high order, you will find that the typical pure mathematician is a sort of Platonist. Only, he is [a] Platonist who corrects the Heraclitan error that the eternal is not continuous. The eternal is for him a world, a cosmos, in which the universe of actual existence is nothing but an arbitrary locus. The end that pure mathematics is pursuing is to discover that real potential world.
>
> (CP 1.646, 1898, *Vitally Important Topics*)

---

[19] CP 1.214, c.1902, *A Detailed Classification of the Sciences*; CP 1.358, c.1890, *A Guess at the Riddle*.

The "real potential world" does not refer to real possibilities and real generals in the sense of Peirce's scholastic realism but to the platonistic, sense-transgressing region of the world of existence. To believe in an eternal but a "real potential world" of mathematical objects waiting to be discovered would be incompatible with the tenets of scholastic realism. According to one of them, general laws are real and operative ingredients of nature. This is not to deny that platonism could not accept change (it does, since forms may undergo change according to Plato's metaphysics), only that: (i) Non-actual but extant forms are determinate in the manner real possibilities are not; (ii) principles accounting for determinate mathematical forms are different from principles operative in Peirce's evolutionary metaphysics; and (iii) in platonism, those principles are not applicable to mathematical truths to the extent they are applicable in pragmaticism.

Second, the platonistic belief in the pre-existing realm of mathematical abstractions is in stark violation of what Peirce dubbed *critical common sensism*, according to which indubitable beliefs, which everyone must possess, are inherently vague and which through criticism such as logical analysis and rational deliberation are subject to revision, change and possible rejection.

3 *Observation and Diagrammatism*

3.1 ICONS AND INDICES IN MATHEMATICS

In not accepting the platonistic conception of the cosmos, Peirce in fact came much closer to Aristotle's philosophy. Concerning experience and observation it is worth noting how he mimicked Aristotle's comment in stating that "nothing emerges in meaningful conception that first does not emerge in perceptual judgment" (EP 2:226, 1903, *Pragmatism as the Logic of Abduction*). He means that all sciences of discovery rest on observation. Observation is either *general* as in philosophy or *singular* as in mathematics and the special sciences, *degenerate* as in mathematics, or *non-degenerate* as in logic, philosophy, and the special sciences. Mathematics in particular "is observational," Peirce explains, "in so far as it makes constructions in the imagination according to abstract precepts, and then observes these imaginary objects, finding in them relations of parts not specified in the precept of construction. This is truly observation, yet certainly in a very peculiar sense; and no other kind of observation would at all answer the purpose of mathematics" (CP 1.240, 1902, *A Detailed Classification of the Sciences*).

On the face of it, this statement does remind us of intuitionism. In it as well as in pragmaticism, mathematical ontology is a hypothetical system of constructs. But pragmatistic constructs are based on very specific kinds of entities: They are *diagrams*. Building upon abstract precepts, mathematicians develop hypotheses based on observing, experiencing, and operating on the relationships exhibited in the diagram and in the domain, including empirical domains. Mathematical objects and structures are thus strictly speaking not abstract even though they are imaginary. What is abstract are the precepts that we employ to oversee the construction of mathematical forms. The diagram is a general but finite schema whose relationships can be predicated of an infinite collection of empirical objects of mathematics. The domains represented by diagrams can be infinitely extendible. This makes Peirce's philosophy quite unlike what Brouwer would have us believe.

The key feature of diagrams is that they are *iconic*: "A great distinguishing property of the Icon is that by the direct observation of it other truths concerning its objects can be discovered than those which suffice to determine its construction" (CP 2.279, 1901, *The Icon, Index and Symbol*). Peirce maintains that iconicity, which often does not involve similarity or likeness—that is, it need not exhibit any "sensual resemblance" (ibid.) between representations and things represented—lies at the bottom of all truly fertile mathematical practices.[20] Icons exhibit abstract and structural relationships and processes that preserve structural properties of the domains of investigation. Peirce provides a number of examples from algebra and geometry to support his notion of iconicity.

---

[20] Peirce states this in the context of his diagrammatic logical theory of existential graphs and sees it indispensable in scientific inquiry: "Diagrammatic reasoning is the only really fertile reasoning. If logicians would only embrace this method, we should no longer see attempts to base their science on the fragile foundations of metaphysics or a psychology not based on logical theory; and there would soon be such an advance in logic that every science would feel the benefit of it" (CP 4.571, 1906, *Prolegomena to an Apology for Pragmaticism*).

In addition to the pivotal role of icons in mathematical reasoning, mathematics needs indexical signs. Since indices are signs that must have objects, mathematical constructs cannot be mere fictions that are not matching up with any mathematical entities. But the observation that concerns the relationships taking place in the creations of our minds is according to Peirce a degenerate form of observation. It means that the objects of such creations need not be parts of the objective reality of the world but may also refer to images created by the previously encountered mathematical signs and their interpretations. But it would be a fallacy to hold degenerate observation to mean imagining objects of fiction consistent with antecedent narratives by mathematicians, since in the diagrammatic structures created by the mind the real relationships are being observed.

Degenerate observation is effectuated by the application of indexical signs. Indices are labels or names attached to diagrams, geometric constructions and different stages of proofs.

> But the imaginary constructions of the mathematician, and even dreams, so far approximate to reality as to have a certain degree of fixity, in consequence of which they can be recognized and identified as individuals. In short, there is a degenerate form of observation which is directed to the creations of our own minds—using the word observation in its full sense as implying some degree of fixity and quasi-reality in the object to which it endeavours to conform. Accordingly, we find that indices are absolutely indispensable in mathematics.
>
> (CP 2.305, 1901, *The Icon, Index and Symbol*)

Aside from the importance of indices in mathematics, the remark concerning the iconic forms of representation in reasoning suggests an explanation for the peculiarity of multiple conclusions in inferential reasoning. It was a problem for the Aristotelian model of scientific reasoning that syllogistics did not seem to leave room for multiple conclusions, though in actual science, such phenomena are commonplace. Those truths that "suffice to determine [the] construction" of the icon are in Peirce's terminology results of *corollarial* deductions, while "other truths" concerning the objects of the icon and the icon's "capacity of revealing unexpected truths" (CP 2.279) refer to *theorematic* deductions.[21] We often fail to see how to derive the latter kinds of conclusions if we try inferring without icons. Experimenting with icons suggests that something needs to be added to the construction of the proof that is not instantiated in the premises. And it is theorematic reasoning that Peirce takes to be the proper kind of reasoning in mathematics; it is "reasoning with specially constructed schemata" (CP 4.233, c.1903, *The Simplest Mathematics*).

Some important links connect these views to areas of modern mathematics. For example, category-theoretic diagrams, such as topoi, are examples of iconic diagrams by which mathematical discovery can be facilitated. If the outcome of the experimentation upon the diagram is that it commutes, we are attributing a novel relational feature to it. This suggests two important issues: (i) that the commutativity of categories is an instance of theorematic reasoning and (ii) that commutativity exemplifies experimental processes by which reasoning proceeds in category-theoretic domains.

Contrary to some suggestions in the recent literature, iconic diagrams are not mere visual aids or heuristics guiding the mathematician into valid conclusions; conclusions that might be reached without such aids as well, though perhaps with more cognitive energy expended. Observations, including direct observations, concern structural similarities between the different domains that diagrams represent. Experimentation enables to expose to view and observe some hitherto unobserved relations. But diagrams need not be visual diagrams. In many instances they are icons that cannot really be visualised. This fits in well in with Peirce's broad conception of a diagram involving also other modes of representation than visual ones and appealing to other types of perception than visual perception. This happens in category theory in which

---

[21] The nature of theorematic reasoning has been extensively discussed in the literature (Hintikka [1980] and Webb [2006]). According to Peirce, "[A] Corollarial Deduction is one which represents the conditions of the conclusion in a diagram and finds from the observation of this diagram, as it is, the truth of the conclusion. A Theorematic Deduction is one which, having represented the conditions of the conclusion in a diagram, performs an ingenious experiment upon the diagram, and by the observation of the diagram, so modified, ascertains the truth of the conclusion" (CP 2.267, 1903, *Division of Signs*).

the iconic relationships have to do with recognitions of abstract and structural likenesses rather than similarities in appearances.

Mathematical objects are not strictly speaking abstract even though they are conceived through imaginary means, and so diagrammatic schemas cannot be held to be abstract objects, either. They are instantiated in concrete systems of relationships that are just as real, although they do not exist in the same way, as physical systems of relationships. Mathematical objects are objects of the signs to which mathematical reasoning as "creations of the mind" refer to, but they fall from the relational structures produced by cognitive processes going on in the mind. Since the mind is not an individual, mathematical objects are not merely our own ideas, although they appear to us as consequences of our cognitive activities.

Diagrammatic approach to mathematics based on iconicity of representations involves experimental reasoning and observation by human mathematicians. It is fallible just as the rest of the sciences are fallible. Although its reasonings are necessary,[22] a closer scrutiny of the methods of establishing reasoning reveals, among other things, that we make experiments upon and observations concerning *continuous* representations. And these continuous representations are iconic diagrams. Since Peirce's continuum contains real uncertainty and vagueness, those methods cannot be relinquished from producing fallible results. The degree of certainty as compared with natural or human sciences may well be higher in mathematics, and its self-correcting processes put in place faster than in other sciences, but it is fallible all the same.

A further case in point concerning the fallible character of mathematics is its hypothetical nature appealing to mental creations organised in diagrammatic images of hypothetical conditionals. Critical common sensism of course still dominates here: We must not begin to doubt anything that we have no reason to doubt.[23]

---

[22] Though as noted, mathematics by no means consists of performing the necessary reasoning.

[23] Cooke ([2003]: 158) argues that the context dependence of science and mathematics in the pragmatic model of inquiry supports the infallibility of mathematics, since inquirers "can never get outside their context of inquiry" to make claims about the certainty of their domains of research. But this view misses the fact that mathematicians and logicians are in fact capable of formulating metamathematical and

3.2 QUASI-EMPIRICISM

A philosophy of mathematics that has gained popularity during the last couple of decades is *quasi-empiricism*.[24] It would be now in order to contrast pragmaticism with quasi-empiricism, too. Hilary Putnam in his commentary to Peirce's *Cambridge Conference Lectures* of 1898 in fact uses the label of quasi-empiricism to tag Peirce's philosophy of mathematics (Peirce [1992]: 74).[25] True, quasi-empiricism shares some of the criticism that Peirce might have been keen to level against the up-and-coming foundational philosophies of mathematics, one of them being the sorting of mathematical truths into factual and conceptual.[26] Quasi-empiricism advances the fallibilistic view that in mathematics, axioms of an axiomatic system may be corrected according to their erroneous consequences and that the laws of logic are subject to change in the light of compelling evidence. It suggests that in mathematics, other methods of reasoning besides deduction are in operation, which of course Peirce recognised long before the coinage of quasi-empiricism.[27]

Given Peirce's attitude that all inquiry is conducted by analogous methods as laboratory activity is conducted, we might, sight unseen, in-

---

metalogical results concerning their domains of research, by approaching those domains piecemeal and by developing new methods to achieve the goals. To "never get outside their context of inquiry" is in fact an anti-pragmatistic claim about the universality of the notion of context (Pietarinen [2007]). After all, such a reliance on the context dependence of inquiry is liable to make all science infallible.

[24] As discussed, among others, by Kitcher ([1984]), Lakatos ([1976]), Maddy ([1997]), Putnam ([1975]), and Quine ([1970]).

[25] The introduction chapter to *Reasoning and the Logic of Things* by Kenneth Laine Ketner and Hilary Putnam also attempts to characterise Peirce's position as an infinitistic version of intuitionism. My paper attempts to show that this cannot be a correct description of Peirce's position, either, since the differences are much more variegated than that of (strict) finitism vs. infinitism.

[26] Better known as the analytic/synthetic distinction. See Levy [1997] and Otte [2006] for related studies. The way in which this dichotomy fails or has been taken to fail is an interesting story of its own.

[27] See e.g. the entry "Logic" in Baldwin's 1901–02 *Dictionary of Philosophy and Psychology* (CP 2.216, 1901, *Why Study Logic?*), in which Peirce laments how the appeal of non-deductive methods of reasoning to mathematics has still not been properly recognised: "The generally received opinion among professors of logic is that all the above methods [of reasoning, namely abduction, deduction, and induction] may properly be used on occasion, the appeal to mathematics, however, being less generally recognized."

terpret Peirce's remarks as embracing the quasi-empiricist vision; that mathematical subjects are indeterminate, that mathematical knowledge is fallible, and that theorems are observation sentences or experimental outcomes of the actual practices and activities of our fellow mathematicians. A quasi-empiricist concludes from these that mathematical and scientific activities are remarkably closely intertwined. The assumption of pragmaticism as a quasi-empiricistic method of investigation in mathematics might also be fuelled by some of Peirce's own remarks, such as one in which he states that "[Mathematics] is supposed to have no empirical element. But this I am quite sure is a serious error" (W4: 556–7, 1884, [*On the Teaching of Mathematics*]; MS 504).

Quasi-empiricism, however, does in one respect not go far enough for Peirce, whose perspective was that all sciences are not only observational but also experimental.[28] The methods of mathematics and the methods of special sciences are continuous in the sense that every experiment is an operation of thought. Branches of sciences are of course observational and experimental in different ways, but in making new discoveries, none of them can dispense with carrying out experiments and then making observations concerning the outcomes of such experiments. The same procedure is, Peirce maintains, in operation in mathematics.

Another realm in which we can discern quasi-empiricist content of mathematical theorising concerns proofs and provability. What counts as a proof and how proofs are carried out depend on the precise nature of experiments performed upon diagrammatic constructions, which are iconic representations of inferential relations holding between premises and conclusions. Here degenerate observation concerns the similarities between the relationships that obtain in diagrams and in what diagrams are representations of. Like many other philosophers of mathematics who followed, Peirce emphasised that the importance of a proof in mathematics lies not in what the proof really is or what the systems

---

[28] Peirce emphasises this at several junctures. Curiously, he also held a separate lecture on the particular issue of observation in mathematics, entitled "The observational element in mathematics." This lecture was presented in a pedagogical series given in 1883–84 at the Johns Hopkins University where Peirce was an instructor in logic at that time (*Johns Hopkins University Circulars*, 3 January 1884: 32). The lecture itself has not been preserved, but MS 748c is a two-page draft of it; cf. W4: 555–8.

or constraints are within which the various proofs can be carried out, but in (i) what it is that the proofs in fact accomplish and (ii) what their general conceivable consequences will be as soon as they do accomplish their purpose.[29]

Another major moot point concerning quasi-empiricism from the vantage point of pragmaticism is its epistemological argument, originally targeted against platonism (Benacerraf [1973] and Benacerraf and Putnam [1983]). According to that argument, we can have mathematical knowledge only of objects we are causally interacting with. I have already suggested how this line of attack remains ineffectual in pragmaticism. It should now be added that, in making Benacerraf-type arguments, it is presupposed that we gain that knowledge through sense perception. That is an assumption pragmaticism does not make. Observation is not mere sense perception. It involves ratiocination.[30] There is no plain and content-free uninterpreted perception given through our senses only. Our senses work much more like "reasoning machines," as Peirce argues in another famous papers of his.[31] And mathematics and logic are both activities of making such observations. If we are willing to grant mathematics the status of one of the higher intellectual activities of humanity, then we should call to mind Peirce's line that, "[T]o say that all our knowledge relates merely to sense perception is to say that we can know nothing—not even mistakenly—about higher matters" (CP 6.492, 1908, *A Neglected Argument for the Reality of God*).[32] No escape from Benacerraf's problem to nominalistic structuralism is needed.

---

[29] On the other hand, quasi-empiricism tends to move too far and into the directions remote from Peirce's concerns, as for instance is the case in the naturalism of Kitcher ([1984]).

[30] According to Peirce, "The investigation of truth consists, according to the conception of logic, of two parts, observation and reasoning. This distinction is not in truth an absolute one. Modern psychology shows us that there is no such thing as pure observation free from reasoning nor as pure reasoning without any observational element" (W4: 400–1, 1883, [*Beginnings of a Logic Book*]).

[31] NEM 3:1115, 1900, *Our Senses as Reasoning Machines*.

[32] Moreover, a sore point in the epistemological argument is its appeal to the notion of causation which by no means is an essential part of pragmatistic methodology.

4 Conclusion

Peirce's pragmaticism is a noteworthy philosophy of mathematics differing markedly from the foundationalist philosophies proposed for mathematics over the course of the century that followed his investigations. But to move away from foundationalist concerns is not to move away from the indubitable yet fallible character of mathematical facts and mathematical knowledge. To do so would contravene the scholastic realist and critical common-sensist views that characterises pragmatistic thought. Pragmaticism is a method of doing philosophy of mathematics by way of drawing from actual mathematical practices and taking what mathematician do seriously.[33]

Acknowledgments

Supported by the University of Helsinki Excellence in Research Grant ("Peirce's Pragmatistic Philosophy and Its Applications"). I presented parts of the paper in the following events: *Perspectives on Mathematical Practices*, Brussels, March 2007; *Logic, Methodology, and the Philosophy of Science*, Beijing, August 2007, and *World Congress of Philosophy*, Seoul, August 2008. My thanks go to all the commentators in these earlier occasions, as well as to the anonymous reader of the earlier version of the present paper.

---

[33] One might suspect that pragmaticism resounds with the kind of natural or naïve attitude towards mathematics taken by Wittgenstein: Mathematics is simply what mathematicians do. But neither Peirce nor Wittgenstein were that casual about the nature of mathematics. According to Peirce, studying the processes of mathematical reasoning that fall from the *logica utens* pertains to the study of normative logic and not to mathematics. Wittgenstein, contra Peirce, took mathematics, too, to be a normative field of investigation. Mathematics is not a science of discovery of new facts, but proving mathematical results means changing the rules of mathematical language games. Overall, however, Wittgenstein held an exceedingly narrow view of what mathematics can accomplish. It is impossible to reconcile Peirce's pragmatistic approach closely tied with actual mathematical investigations with Wittgenstein's finitistic and normative constructivism. Though differing with respect to the philosophy of mathematics, I have investigated some commonalities between Peirce's and Wittgenstein's philosophies of language and logic in Pietarinen [2003] and [2006a].

Abbreviations

CP	Peirce [1958]	EP	Peirce [1998]	MS	Peirce [1967]
NEM	Peirce [1976]	W	Peirce [1980]		

Bibliography

Bell, J.L. [1988]. *Toposes and Local Set Theories.* Oxford: Oxford University Press.

Benacerraf, P. [1973]. "Mathematical truth." *Journal of Philosophy* **70**. 661–80.

Benacerraf, P. and H. Putnam, eds. [1983]. *Philosophy of Mathematics.* 2nd ed. Cambridge: Cambridge University Press.

Blais, M.J. [1989]. "A pragmatic analysis of mathematical realism and intuitionism." *Philosophica Mathematica* **2**. 61–85.

Brouwer, L.E.J. [1975]. "On the foundations of mathematics." *L.E.J. Brouwer, Collected Works.* Ed. by A. Heyting. Vol. 1. Philosophy and Foundations of Mathematics. First published in 1907. Amsterdam: North-Holland.

Burgess, J.P. [2004]. "Mathematics and Bleak House." *Philosophia Mathematica* **12**. 18–36.

Colyvan, M. [2001]. *The Indispensability of Mathematics.* Oxford: Oxford University Press.

Cooke, E.F. [2003]. "Peirce, fallibilism, and the science of mathematics." *Philosophia Mathematica* **11**. 158–75.

Ehrlich, P. [forthcoming]. "The absolute arithmetic continuum and its Peircean counterpart." *New Essays on the Mathematical Philosophy of C. S. Peirce.* Ed. by M. Moore. Open Court.

Engel-Tiercelin, C. [1993]. "Peirce's realistic approach to mathematic: Or, can one be a realist without being a Platonist?" *Charles S. Peirce and the Philosophy of Science: Papers from the Harvard Sesquicentennial Congress.* Ed. by E.C. Moore. Tuscaloosa: University of Alabama Press. 30–48.

Field, H. [1980]. *Science Without Numbers: A Defence of Nominalism.* Oxford: Blackwell.

— [1989]. *Realism, Mathematics & Modality.* Oxford: Blackwell.

Hellman, G. [1989]. *Mathematics without Numbers: Towards a Modal-Structural Interpretation.* Oxford: Clarendon Press.

— [2004]. "Does category theory provide a framework for mathematical structuralism?" *Philosophia Mathematica* **12**. 129–57.

Hintikka, J. [1980]. "C. S. Peirce's 'first real discovery' and its contemporary relevance." *The Monist* **63**. 304–13.

Hoffman, S. [2004]. "Kitcher, ideal agents, and fictionalism." *Philosophia Mathematica* **12**. 3–17.

Hudry, J.-L. [2004]. "Peirce's potential continuity and pure geometry." *Transactions of the Charles S. Peirce Society: A Quarterly Journal in American Philosophy* **40**. 229–44.

Kitcher, P. [1984]. *The Nature of Mathematical Knowledge.* Oxford: Oxford University Press.

Lakatos, I. [1976]. *Proofs and Refutations: The Logic of Mathematical Discovery.* Cambridge: Cambridge University Press.

Levy, S.H. [1997]. "Peirce's theoremic/corollarial distinction and the interconnections between mathematics and logic." Ed. by N. Houser; D.D. Roberts; and J. Van Evra. *Studies in the Logic of Charles S. Peirce.* 85–110.

Maddy, P. [1997]. *Naturalism in Mathematics.* Oxford: Clarendon Press.

Murphey, M.G. [1961]. *The Development of Peirce's Philosophy.* Cambridge, MA: Harvard University Press.

Myrwold, W.C. [1995]. "Peirce on Cantor's paradox and the continuum." *Transactions of the Charles C.S. Peirce Society: A Quarterly Journal in American Philosophy* **31**. 508–41.

Otte, M. [2006]. "The analytic/synthetic distinction and Peirce's conception of mathematics." *Semiotics and Philosophy in Charles Sanders Peirce.* Ed. by R. Fabbrichesi and S. Marietti. Cambridge: Cambridge Scholars Press. 51–88.

Patin, H.A. [1957]. "Pragmatism, intuitionism, and formalism." *Philosophy of Science* **24**. 243–52.

Peirce, C.S. [1958]. *Collected Papers of Charles Sanders Peirce.* 1931–58. Ed. by C. Hartshorne; P. Weiss; and A.W. Burks. Cambridge, MA: Harvard University Press.

Peirce, C.S. [1967]. *Manuscripts in the Houghton Library of Harvard University.* Identified by Richard Robin. See also Peirce [1971]. Amherst: University of Massachusetts Press.

— [1971]. "Manuscripts in the Houghton Library of Harvard University." *The Peirce Papers: A supplementary catalogue.* Vol. 7. Transactions of the C. S. Peirce Society. 37–57.

— [1976]. *The New Elements of Mathematics.* Ed. by C. Eisele. The Hague: Mouton Publishers.

— [1980]. *The Writings of Charles S. Peirce.* Ed. by M. Fisch et al. Bloomington: Indiana University Press.

— [1992]. *Reasoning and the Logic of Things: The Cambridge Conference Lectures of 1898.* Ed. by K. Ketner and H. Putnam. Harvard: Harvard University Press.

— [1998]. *The Essential Peirce. Selected Philosophical Writings (1893–1913).* Ed. by The Peirce Edition Project. Vol. 2. Bloomington and Indianapolis: Indiana University Press.

Pietarinen, A.-V. [2003]. "Logic, language games and ludics." *Acta Analytica* **18**. 89–123.

— [2005]. "Cultivating habits of reason: Peirce and the *logica utens* vs. *logic docens* distinction." *History of Philosophy Quarterly* **22**. 357–72.

— [2006a]. "Interdisciplinarity and Peirce's classification of the sciences: A centennial reassessment." *Perspectives on Science* **14**. 127–52.

— [2006b]. *Signs of Logic: Peircean Themes on the Philosophy of Language Games, and Communication.* Vol. 329. Synthese Library. Dordrecht: Springer.

— [2007]. "Semantics/pragmatics distinction from the game-theoretic point of view." *Game Theory and Linguistic Meaning.* Ed. by Ahti-Veikko Pietarinen. Vol. 18. Current Research in the Semantics/Pragmatics Interface. Oxford: Elsevier. 229–42.

— [2008]. "The Proof of Pragmatism: Comments on Chris Hookway." *Cognitio* **9**. 85–92.

— [forthcoming]. "Which philosophy of mathematics is pragmaticism?" *New Essays on the Mathematical Philosophy of C. S. Peirce.* Ed. by Matthew Moore. Open Court.

Putnam, H. [1975]. *Mathematics, Matter and Method*. Cambridge: Cambridge University Press.
Quine, W.V.O. [1970]. *Philosophy of Logic*. 2nd ed. Cambridge, MA: Harvard University Press.
Shapiro, S. [1997]. *Philosophy of Mathematics: Structure and Ontology*. Oxford: Oxford University Press.
Stjernfelt, F. [2007]. *Diagrammatology: An Investigation on the Borderlines of Phenomenology, Ontology, and Semiotics*. Dordrecht: Springer.
Thomas, R.S.D. [2007]. "The comparison of mathematics with narrative." *Perspectives of Mathematical Practices*. Ed. by B. Van Kerkhove and J.P. Van Bendegem. Dordrecht: Springer. 43–59.
Webb, J. [2006]. "Hintikka on Aristotelian constructions, Kantian intuitions, and Peircean theorems." *The Philosophy of Jaakko Hintikka*. Ed. by R.E. Auxier and L.E. Hahn. Library of Living Philosophers. Chicago and La Salle: Open Court. 195–301.
Weyl, H. [1918]. *Das Kontinuum*. Leipzig: Veit.
— [1987]. *The Continuum: A Critical Examination of the Foundation of Analysis*. Trans. by S. Pollard and T. Bole. First published in German as Weyl [1918]. Kirksville. Mo.: Thomas Jefferson University Press.

ESSAY 8

Mathematical Knowledge as a Case Study in Empirical Philosophy of Mathematics

Benedikt Löwe
Thomas Müller
Eva Müller-Hill

Empirical research in philosophy of mathematics is rare. Possibly, this can be traced back to the (Kantian) conviction that mathematics does not depend on human experience, but is accessible to the *reine Vernunft*. According to this conviction, the truth of mathematical statements and the status of mathematical knowledge should not depend on contingent facts about humans or human society. This view has not only left traces in philosophy of mathematics, but can also be seen in the practice of sociology: Compared to sociological research dealing with other sciences, sociology of mathematics is severely underrepresented.[1]

On the other hand, some of the central questions of philosophy of mathematics have an empirical core, and some of the statements that one finds in philosophical texts about mathematics are empirical claims.

For instance, consider the question of the connection between philosophical views and mathematical achievements. According to mainstream philosophical belief, many working mathematicians are Platonists believing in the actual existence of mathematical objects. It is part of this mainstream belief that these mathematicians interpret their own work as mental manipulation of abstract objects or of mental representations of abstract objects. This last claim is without any doubt an empirical claim about mathematicians. It is an interesting question whether there is a correlation between philosophical attitudes and mathematical success, or—to put it more bluntly—whether you have to be a Platonist in order to be a great mathematician. This statement can be true or false; the reasons for its truth or falsity need not be empirical; but the statement itself is an empirical statement, and the most natural way to find out whether it is true or false is to test it. Very few philosophers of mathematics take this last step, and, as mentioned before, it is not an easy step to take, as data on these questions is not abundant.

Philosophy of mathematics, like other areas of philosophy, relates phenomena (in this case, mathematics) to a philosophical theory. Whether the philosophical theory is correct/adequate or not is not independent of the phenomena. In analytic philosophy and in particular in philosophical logic, the analysis of phenomena is often done by a technique that one could call *conceptual modelling*, *philosophical modelling*, or *logical modelling*, in analogy to the well-known applied mathematics technique of mathematical modelling. This technique consists of a number of natural steps, one of which is to confront the philosophical model with the phenomena. We claim that in many areas of philosophy, especially in the case of philosophy of mathematics, this step is under-

---

[1]Cf. Heintz [2000]: 9, "Die Soziologie [begegnet] der Mathematik mit einer eigentümlichen Mischung aus Devotion und Desinteresse" (sociology meets mathematics with a queer mixture of devotion and lack of interest). Her study thus reconfirms the earlier assessment of Latour ([1987]: 245f).

developed, and we propose to consider collecting data that allow us to identify stable philosophical phenomena in mathematical practice.

This paper is a case study that should be seen as a part of a much larger research agenda of the first two authors: the development of an empirical study of philosophy of mathematics. Such an approach could be called "naturalistic" (as in Maddy [1997]) or a "Second Philosophy of Mathematics" (as in Maddy [2007]). We will use the label "*Empirical Philosophy of Mathematics*" in order to stress the fact that there is actual empirical work to be done in this field. The project *Empirical Philosophy of Mathematics* consists of a theoretical foundation and an unlimited number of questions and practical projects. Some first steps towards *Empirical Philosophy of Mathematics* have been documented in Löwe and Müller [2008]; Müller-Hill [2009]; Löwe [2007]. The theoretical foundation should contain a sustained argument for the methodology of conceptual modelling as pointed to above, especially an argument for the necessity of empirical checks on the philosophical theories established via this method. This argument is not meant to operate on the (probably hopeless) level of generality of "giving a methodological basis for philosophy," but rather should address the specific methodological issues of one particular argumentation scheme in philosophy, in our case, exemplified in philosophy of mathematics. Practical projects in *Empirical Philosophy of Mathematics* might, e.g., address a single philosophically interesting question about mathematics and employ empirical methods from psychology, cognitive science, linguistics, textual analysis, and quantitative or qualitative sociology, to provide an empirically well-founded answer to the given question.

In accord with what has been said above, this paper has two parts: a discussion of the method of *Empirical Philosophy of Mathematics*, and a case study addressing one specific aspect of the analysis of knowledge ascriptions in mathematics. In the first part, we shall mostly proceed descriptively, with few examples and no arguments.[2] Section 1 gives some background on the practice of mathematical modelling, and Section 2 draws parallels to the case of philosophical, or conceptual, modelling.

---

[2]For a further discussion of the motivation for a new approach to philosophy of mathematics, we refer the reader to the introductory paper Buldt, Löwe, and Müller [2008] in the special issue of the journal *Erkenntnis* entitled "Towards a New Epistemology of Mathematics."

However, the proper development of the main tenets of the theoretical foundations for *Empirical Philosophy of Mathematics* has to be left for future work. In the second part (Section 3), we use data procured by the third author, described in more detail in the paper Müller-Hill [2009], to answer a specific question in the epistemology of mathematics. This part is meant to exemplify how the practical side of *Empirical Philosophy of Mathematics* would be carried out. We trust that this will provide enough detail to get an impression of the interplay between philosophical work and empirical studies.

1 Mathematical Modelling

The notion of a model has acquired a prominent place in contemporary philosophy of science. A great variety of uses of the term "model" has been studied (cf. Frigg and Hartmann [2006] for an overview). There is widespread agreement that models play a crucial role in scientific practice, and that a fair amount of that practice consists in modelling. In Section 2 we will suggest that the modelling paradigm also provides a good framework for certain philosophical investigations, and we will propose a modelling paradigm, which could be called *philosophical modelling* or *conceptual modelling*, as an approach to philosophy of mathematics.

In order to have some basis for drawing an analogy, in this section we will describe the practice of mathematical modelling, as exemplified in the sciences. The aim of this exposition is not to give an in-depth account of mathematical modelling, but to show its features relevant for the discussion in Section 2.

Following the Galilean conviction that the book of nature è *scritto in lingua matematica*, in science it is customary to employ mathematical methods to describe, explain, or predict phenomena wherever possible. Kepler's work on planetary motion illustrates important aspects of such mathematical modelling. Kepler first proposed a model of the solar system in which a number of spheres containing the planets were arranged in an intricate manner determined by the five Platonic solids. This model was inspired by the conviction that God's universe must exhibit a high degree of mathematical harmony, and it did fit the then available data reasonably well. Later on, when Tycho Brahe's more accurate data became available to him, Kepler proposed different mathematical models containing what we now call Kepler's three laws of planetary motion, describing elliptical instead of spherical orbits of the planets.

With hindsight, one might describe this episode in the history of science as follows:

(1) Kepler started with a view of what a model of the solar system should look like. His view was that God must have arranged the (then known) six planets in such a way that their spatial layout is determined by some order of the five Platonic solids.

(2) Guided by the available (Copernican) data, he found the best model from the mentioned class, i.e., a particular ordering of the solids.

(3) A confrontation with Brahe's data afterwards showed that the model was inadequate. Kepler began to look for a new class of models, still believing that there must be mathematical harmony in the motion of the planets.

(4) Thus Kepler went through a new cycle of modelling. He tried out a number of algebraic descriptions of planetary motion and finally arrived at his celebrated laws.

Abstracting from the historical setting, one can formulate an iterative method of mathematical modelling exhibited in this, but also in many other episodes from science:

Step 1 One starts with a class of models that appear as reasonable candidates. This class may be determined by pre-theoretical insight, or by earlier steps in the iteration.

Step 2 One collects data and tries to achieve a best fit within the available class of models.

Step 3 One determines the goodness of fit and will either be satisfied or revert to Step 1, having chosen a different class of models.

This basic scheme is at work in many areas of science, both historically and at present. Statistical tools have been developed for assessing the "goodness of fit" of models and data. There is usually an additional layer of modelling for the data themselves (in order to handle measurement errors), and the determination of best-fit models is usually carried out on a computer (cf., e.g., Gershenfeld [1999]). But the underlying structure appears to be preserved.

In mathematical modelling, the step of confronting the selected model with the observed phenomena is absolutely crucial. As every scientist will be proud to say, honesty with respect to that step is the mark of good science: A beautiful mathematical theory is worth nothing if it doesn't fit the phenomena in question. Whether this ideal is always achieved in practice is, of course, another question, but the overall methodology is clear: Modelling is an iterative procedure, and unless one has achieved a good fit, one cannot leave the circle.

2 Conceptual Modelling

Viewed abstractly, the aim of establishing a "philosophy of X" is quite similar to finding a "model for Y" in the sciences: One wishes to gain theoretical insight into (some) aspects of a certain phenomenon by representing them in a specific way.[3] In epistemology, to give just one example, one of the key questions is what knowledge consists in. Various models of knowledge are on the market; a time-honored conception, dating back to Plato's *Meno*, takes knowledge to be justified true belief.

In an episode resembling Kepler's confrontation of his Platonic model of the heavens with Brahe's observational data, the Platonic conception of knowledge was also challenged by data taking the form of counterexamples: Gettier ([1963]) constructed plausible scenarios in which persons have justified true belief, but not knowledge. (E.g., someone might believe in a true disjunction, but her justification supports the false disjunct, while the other is, unbeknownst to her, in fact true.) The ensuing debate led to a repertoire of test cases that serves as a benchmark for theories of knowledge: One way or another, any respectable analytic model of knowledge has to answer the challenges posed by Gettier- and post-Gettier-examples.

As in the scientific case described in the previous section, the actual

---

[3] The question whether that is what's behind the demand for an "explanation" of the phenomenon must be left open at this point.

details of the story are of course more complicated. Surely everybody agrees that if Gettier cases are cases of justified true belief without knowledge, then knowledge cannot be justified true belief. But how do we know? It may be reasonably simple to determine whether a person has a justified true belief (depending on one's account of justification)— but how to judge the presence or absence of knowledge in a given scenario? People certainly have their intuitions and gut feelings, and in many cases that may be sufficient. But in the epistemological case at hand, intuition confronts intuition.

It is for such cases that we propose to mirror the method of mathematical modelling in the sciences more closely in a philosophical context.[4] Thus, conceptual modelling of X also takes the form of an iterative process:

Step 1. Theory formation Guided by either a pretheoretic understanding of X or the earlier steps in the iteration, one develops a structural philosophical account of X, including, e.g., considerations of ontology and epistemology.

Step 2. Phenomenology With a view towards Step 3, one collects data about X and extracts stable phenomena from them that are able either to corroborate or to question the current theory.

Step 3. Reflection In a circle between the philosophical theory, the philosophical theory formation process and the phenomenology, one assesses the adequacy of the theory and potentially revises the theory by reverting to Step 1.

In many debates of contemporary epistemology, Step 2 consists of a presentation of the author's intuitions about the case at hand, possibly supported by anecdotal evidence. While this may be enough if there is widespread consensus about the analysis, a different solution needs to be found if one has to decide between competing models. The obvious solution, in view of the scientific modelling practice, is to supply more data from a more varied range of sources, including data established via accepted empirical methods. On this view, the key to successful conceptual modelling lies in strengthening Step 2 of the above iterative

---

[4]For more details, cf. Löwe and Müller [to appear].

scheme.⁵ In epistemology, the necessary data might, e.g., be supplied from empirical linguistics or from cognitive science. In the recent epistemological debate the need for such an empirical component is felt rather strongly, and a number of authors have discussed the philosophical relevance of "going empirical."⁶

Similarly, philosophers of mind might in their investigations refer to results from cognitive science or neuroscience, while philosophers of language might refer to corpora from linguistics. Stich ([2001]) has argued that quite generally, it is time for philosophy to stop relying on intuitions, i.e., translated into our terminology: to look for an empirical basis for Step 2. In the philosophy of a science X, the relevant data might come from a description of the scientific practice, e.g., as provided by an empirically founded sociology of X. Since the 1980s, such an approach to philosophy of physics and biology has proved fruitful, even though there is no universal consensus about all of the results (Hacking [1999]; Harding [1986]; Knorr Cetina [1999]; Latour and Woolgar [1979]; and Pickering [1984]).

In a similar vein, many authors emphasize the social embedding of mathematics as an important factor for philosophy of mathematics, for instance, the late Wittgenstein, Philip Kitcher ([1984]), Paul Ernest ([1998]), or David Bloor ([1996] and [2004]). Yet, there has hardly been done any *empirical* research about actual mathematical research practice. As mentioned in the introduction, sociology of science has, with few exceptions, shunned away from taking mathematics as an object of study—mainly just because preconceived philosophical convictions made such studies appear senseless or impossible (cf. Footnote 1). The first large-scale socio-empirical study published was Bettina Heintz's

---

⁵In this vein it is easier to reach a negative verdict (stop at Step 3) than to establish a positive result (start afresh at Step 1 and complete the cycle successfully). This may explain the affinity of analytic philosophy to producing more negative than positive results. Cf. also Löwe and Müller [to appear]: §5.3. This affinity is connected to Gian-Carlo Rota's ([1991]) sharp criticism of what he calls "mathematical philosophy."

⁶For a still rather traditional example of employing linguistic data in epistemology, cf. Stanley [2004]. More decidedly experimental views are discussed in Weinberg, Nichols, and Stich [2001]; Alexander and Weinberg [2006]; Nagel [2007]; Swain, Alexander, and Weinberg [2008]. For a detailed discussion of this debate in the framework of conceptual modelling, cf. Löwe and Müller [to appear]: §4.3.

([2000]) work about the culture and practice of mathematics as a scientific discipline mentioned in the introduction.[7]

In philosophy of mathematics, conceptual modelling according to the above scheme is therefore hampered by the fact that there is little empirical data available; accordingly, one has to work on Step 2 and Step 3 of the modelling cycle: Acquire data and develop instruments to interpret these data. The objective of *Empirical Philosophy of Mathematics* is to integrate these two objectives, working hand in hand with sociologists, cognitive scientists, psychologists, and experts in mathematics education in order to gain access to the necessary empirical data. This is not an easy task, and it is only too tempting to oversteer: While trying to emphasize the importance of Step 2, one could forget the original philosophical agenda and transform *Empirical Philosophy of Mathematics* into pure sociology of mathematics. The philosopher has to ask herself what the philosophical content of empirical facts such as "the majority of mathematicians accepts that X knows that p" is. How do we transcend the mere description of what mathematicians are doing and enter the study of understanding what mathematics is? In some cases, the empirical findings, or even the experience of collecting the empirical data, will be philosophically relevant for the argumentation for or against the particular methodology employed. This emphasizes the close relationship of the modelling cycle of conceptual modelling as described above and the hermeneutic circle.

Since giving an exhaustive description of and argument for the methodology of conceptual modelling is impossible, we will illustrate this approach in Section 3 via a case study that employed methods from quantitative sociology to generate the data. We think that this explains the methodology (and its difficulties) more than any attempt of a theoretical description. In our case study, the aim is to incorporate empirical data about actual mathematical practice in order to elucidate the knowledge concept used in mathematical practice. This should provide input for a philosophical theory of mathematical knowledge and the epistemological role of formalizability in mathematics. In terms of our modelling scheme, the study proceeds from Step 1 to Step 3, reaching a

---

[7]Before Heintz ([2000]), Markowitsch used qualitative sociological studies (interviews with mathematicians) in his [1997].

negative conclusion about the model chosen in Step 1.

3 Case Study: The Role of Formal Proofs in Mathematical Epistemology

3.1 THE PHILOSOPHICAL ISSUE AND THE EMPIRICAL TEST QUESTIONS

Mathematical knowledge is generally assumed to be absolute and undeniably firm. The main reason for that special status lies in the fact that mathematicians *prove* their theorems: Mathematical knowledge is proven knowledge (*"more geometrico demonstrata"*). Thus, mathematical knowledge stands out as knowledge with a uniform witness, the notion of proof, which since the end of the 19th century has been connected to the ideal of formal proof. Mathematical certainty seems to come from the fact that we believe that it is possible to formalize all informal proofs.

Even though this standard view of knowledge in mathematics is central in philosophical debates, neither formal proof nor the formalizations of informal proofs play a big role in actual mathematical practice. The philosopher of mathematics, at least if she wishes to develop a philosophy of mathematics that reflects to a certain extent mathematical practice, should try to explain this discrepancy and answer the question: What is essential about formalizability for a philosophical understanding of *mathematical knowledge*?

The traditional view could for example be presented as follows: For a mathematician X and a mathematical statement φ, "X knows that φ" holds if and only if X has access to an informal proof of φ that can be rendered into a formal derivation.

Note that such a characterization of mathematical knowledge is not a precise philosophical position since various notions ("has access to," "can be rendered") are vague and need explication.[8] However, even from this imprecise characterization, one can deduce a number of features of mathematical knowledge (under very mild assumptions about the vague terms in the characterization). For instance, unless there is an

---

[8]This is discussed in more detail in Löwe and Müller [2008]: §3.

inconsistency in our basic axioms of mathematics, it would be impossible for X to know φ at time t and to know $\neg\varphi$ at time t', even if $t \neq t'$: If X had access to an informal proof at time t that can be transformed into a derivation D of φ and had access to an informal proof at time t' that can be transformed into a derivation D' of $\neg\varphi$, then the derivations D and D' witness that there must be an inconsistency in our underlying base axiom system.[9]

In the naturalistic spirit of *Empirical Philosophy of Mathematics*, we can test the above traditional view of mathematical knowledge: If the usage of mathematicians allows for conflicting knowledge ascriptions, then the traditional view must be abandoned.

It should be emphasized that it is an interesting question whether the results of our investigation are specific for mathematics at all. Our report here focuses on one particular aspect of the questionnaire study that was conducted, and could give the impression that the only thing that we are interested in is the question "Is the actual usage of mathematical knowledge attributions compatible with facticity of knowledge?" (or, more specifically, "Can you adequately/truthfully say 'I knew it this morning, but now I don't know it anymore'?"). Reduced to this question, you might wonder whether our investigation would not yield exactly the same results if you told a story about biologists, physicists, or just ordinary people talking about red cars or clocks. And it actually might. We think that this is an interesting empirical research question for linguists with relevance for epistemology. If anything, such an empir-

---

[9] As one referee pointed out, this is assuming that there are no "different acceptable axiomatizations at different times." This description can refer to several phenomena in mathematics, related to the subtle gradation of the use of axiomatics: On one end of the spectrum, axioms serve as part of the definition of a concept, on the other end, they are self-evident truths about mathematics. In the first case, "different acceptable axiomatizations" might refer to historical examples of changes in the meaning of basic concepts (cf. Buldt, Löwe, and Müller [2008]: §2 for a discussion of the notion of continuity); in the second case, we would be talking about something that Maddy ([1997]: 208) would call a failure of the maxim UNIFY. Both phenomena are interesting and worthy of study within our paradigm of *Empirical Philosophy of Mathematics*, but for the current investigation we will disregard the possibility of such a situation—in the example employed in our case study, a change in axiomatization is not an issue.

ical finding would only corroborate our belief that the idea of a special epistemic status of mathematics is highly problematic.

3.2 OUR TEST

In order to test the traditional view that mathematical knowledge is special because it is based on formalizable proof, we set up a quantitative experiment. It is difficult to get access to large enough numbers of test subjects for questions like this[10]; for this study, we used an online questionnaire that was advertised via scientific newsgroups. This guaranteed a relatively large number of test subjects, but limited the study to mathematicians who read newsgroups. Yet, we do not regard this as a serious limitation of the representativity of the sample, as a significant correlation between the habit of reading newsgroups and a certain attitude towards formalization is not very likely.

Our questionnaire contained mostly multiple choice questions, but also left some space for qualitative comments. It was posted on a webpage from August 2006 to October 2006 and announced via postings in three different scientific newsgroups, with about 100 valid responses out of 250 responses in total. A response was considered as valid if the personal data part and at least one question of the questionnaire was completed.[11]

The personal data part served to decide whether a participant belonged to the target group of participants who have been involved in research or university-level teaching. We got 76 valid responses from the target group. Most of these responses came from the US, Germany, and the Netherlands. 46.1% of the target group participants hold a Ph.D. (or an equivalent degree), 19.7% an M.Sc. (or an equivalent degree), and 13.2% a B.Sc. (or an equivalent degree) in mathematics.

3.3 SELECTED RESULTS

The questionnaire had 74 questions in total, but we will only present some questions and results that are significant for the interpretation

---

[10]Together with Francesco Lanzillotti and Henrik Nordmark, the first author has attempted to access research mathematicians via the organizers of international conferences with extremely limited success.

[11]In the presentation of the results in Section 3.3 we will give the total count of valid answers for each reported multiple choice question.

regarding the test question.[12]

In one part of the questionnaire, participants were led through four scenarios. Each screen of the online questionnaire contained a piece of the scenario, and at the end of each screen the participants were asked whether they would ascribe knowledge to the protagonist of the scenario or not. In the following, we will give excerpts from one of these scenarios.

> John is a graduate student, and Jane Jones, a world famous expert on holomorphic functions, is his supervisor. One evening, John is working on the Jones conjecture and seems to have made a break-through. He produces scribbled notes on yellow sheets of paper and convinces himself that these notes constitute a proof of his theorem.
>
> Does John know that the Jones conjecture is true?
> ☐ Yes
> ☐ Almost surely yes
> ☐ Almost surely no
> ☐ No
> ☐ Can't tell

Screenshot 1: First screen of the "Jones Conjecture" scenario

On the following screens, the story continues: John presents the proof sketch to his professor, she discovers a gap and fixes it, they jointly write a paper that they submit to a mathematical journal of high reputation. After a thorough refereeing process of 18 months, the paper is accepted on the basis of a positive referee report. At the end of each screen the participants are asked the same question, "Does John know that the Jones conjecture is true?" Finally, the scenario ends with Screenshot 2.

Table 1 contains the results for three of the questions for the Jones scenario. Q1 is the question whether John knows that the Jones conjecture is *true* after the paper was published, Q2 is the question whether John knows that the Jones conjecture is *false* after he discovers the coun-

---
[12]Cf. Müller-Hill [2009] for a more detailed exposition of the results of the survey.

> After his Ph.D., John continues his mathematical career. Five years after the paper was published, he listens to a talk on anti-Jones functions. That evening, he discovers that based on these functions, one can construct a counterexample to the Jones conjecture. He is shocked, and so is professor Jones.
>
> Does John know that the Jones conjecture is false?
>
> ☐ Yes
> ☐ Almost surely yes
> ☐ Almost surely no
> ☐ No
> ☐ Can't tell
>
> Did John know that the Jones conjecture was true on the morning before the talk?
>
> ☐ Yes
> ☐ Almost surely yes
> ☐ Almost surely no
> ☐ No
> ☐ Can't tell

Screenshot 2: Final screen of the "Jones Conjecture" scenario

terexample, and Q3 is the question whether he knew that the conjecture was *true* on the morning before the talk. Note that Q2 and Q3 are on the same screen (see above), so participants are aware of their answer to Q2 when answering Q3 and vice versa. After finishing the scenario, the participants were given the opportunity to write down comments on the scenario in a free-text field. Here are some selected quotes, emphasizing different factors that influenced the answers in the Jones scenario:

- "How important is the Jones conjecture? How large is the community?"

- "My answers would have been very different with different time frames mentioned."

		Yes	Almost surely yes	Almost surely no	No	Can't tell
Q1	Count Σ = 66	19	37	2	3	5
	Frequency	28.8%	56.1%	3.0%	4.5%	7.6%
Q2	Count Σ = 62	9	29	4	5	15
	Frequency	14.5%	46.8%	6.4%	8.1%	24.2%
Q3	Count Σ = 62	15	29	3	9	6
	Frequency	24.2%	46.8%	4.8%	14.5%	9.7%

Table 1: Results for the Jones scenario

- "I don't know John . . . , so I don't have a good feel for how rigorously [he] . . . work[s]."

3.4 INTERPRETATION

The results from the Jones scenario are to a certain extent as one would expect. After thorough refereeing and publication of the result in a respected journal, John is taken to *know* that the Jones conjecture is true—the great majority (84.9%) of the participants gave a positive answer ("yes" or "almost surely yes") to Q1. After discovery of the counterexample, a majority ascribes to John knowledge that the Jones conjecture is *false*—61.3% of the participants gave a positive answer to Q2, and only 14.6% deny his knowledge ("no" or "almost surely no"). The answers to Q3, however, might be surprising: In a situation in which the storyline has revealed that there is a counterexample to the conjecture, still a majority of 71% gave a positive answer to that question, i.e., would ascribe knowledge of the conjecture on the morning before he learned about the counterexample.

In order to appreciate this phenomenon, let us look at the correlation figures. Of the 38 participants who agreed to Q2, 27 also agreed to Q3 (71.1%). With this empirical result, we can now move to Step 3 of the modelling cycle: As argued in Section 3.1, a philosopher endorsing the traditional view of epistemology of mathematics is not able to accept the statement "John knows φ at time t and John knows $\neg\varphi$ at time t'." However, a large number of our test subjects did exactly that, and so we either have to accept a notion of mathematical knowledge that ignores the usage of this large portion of the community, or give up the traditional view. Based on our methodological position, we discard the former option and choose the latter. The free text comments given in the survey further support the view that mathematical knowledge is highly context-sensitive.

Our socio-empirical analysis gives a negative result; it is now our task as philosophers of mathematics to turn it into a philosophical insight: In the modelling cycle, we are now returning to Step 1. Based on the quantitative data, the free text comments, and further results from a qualitative interview study with mathematicians (planned by the third author), we should now develop a new understanding of knowledge that replaces the one that we called the "traditional view."

Bibliography

Alexander, J. and J.M. Weinberg [2006]. "Analytic Epistemology and Experimental Philosophy." *Philosophy Compass* **1**. 56–80.

Bloor, D. [1996]. *Scientific Knowledge: A Sociological Analysis*. Chicago, IL: University of Chicago Press.

— [2004]. "Sociology of Scientific Knowledge." *Handbook of Epistemology*. Ed. by I. Niiniluoto; M. Sintonen; and J. Woleński. Dordrecht: Kluwer. 919–62.

Buldt, B.; B. Löwe; and T. Müller [2008]. "Towards a New Epistomology of Mathematics." *Erkenntnis* **68** (3). 309–29.

Ernest, P. [1998]. *Social Constructivism as a Philosophy of Mathematics*. Albany, NY: SUNY Press.

Frigg, R. and S. Hartmann [2006]. "Models in Science." *Stanford Encyclopedia of Philosophy*. Ed. by E.N. Zalta. Spring 2006 edition. CSLI.

Gershenfeld, N. [1999]. *The Nature of Mathematical Modelling*. Cambridge: Cambridge University Press.

Gettier, E. [1963]. "Is Justified True Belief Knowledge?" *Analysis* **23**. 121–3.

Hacking, I. [1999]. *The Social Construction of What?* Cambridge, MA: Harvard University Press.

Harding, S. [1986]. *The Science Question in Feminism*. Ithaca, NY: Cornell University Press.

Heintz, B. [2000]. *Die Innenwelt der Mathematik. Zur Kultur und Praxis einer beweisenden Disziplin*. Wien: Springer.

Kitcher, P. [1984]. *The Nature of Mathematical Knowledge*. Oxford: Oxford University Press.

Knorr Cetina, K. [1999]. *Epistemic Cultures*. Cambridge, MA: Harvard University Press.

Latour, B. [1987]. *Science in Action. How to follow scientists and engineers through society*. Cambridge, MA: Harvard University Press.

Latour, B. and S. Woolgar [1979]. *Laboratory Life. The Social Conatruction of Scientific Facts*. London: Sage.
Löwe, B. [2007]. "Visualization of ordinals." *Logik, Begriffe, Prinzipien des Handelns*. Ed. by T. Müller and A. Newen. Paderborn: Mentis. 64–80.
Löwe, B. and T. Müller [2008]. "Mathematical Knowledge is Context-Dependent." *Grazer Philosophische Studien* **76**. 91–107.
— [to appear]. "Data and phenomena in conceptual modelling." *Synthese*.
Löwe, B.; V. Peckhaus; and T. Räsch, eds. [2006]. *Foundations of the Formal Sciences IV: The History of the Concept of the Formal Sciences*. Vol. 3. Studies in Logic. London: College Publications.
Maddy, P. [1997]. *Naturalism in Mathematics*. Oxford: Oxford University Press.
— [2007]. *Second Philosophy: A Naturalistic Method*. Oxford: Oxford University Press.
Markowitsch, J. [1997]. "Metaphysik und Mathematik. Über implizites Wissen, Verstehen und die Praxis in der Mathematik." PhD thesis. Universität Wien.
Müller-Hill, E. [2009]. "Formalizability and Knowledge Ascriptions in Mathematical Practice." *Philosophia Scientiae* **13** (2). 21–43.
Nagel, J. [2007]. "Epistemic Intuitions." *Philosophy Compass* **2**. 792–819.
Pickering, A. [1984]. *Constructing Quarks*. Chicago: University of Chicago Press.
Rota, G.-C. [1991]. "The pernicious influence of mathematics upon philosophy." *Synthese* **88**. 165–78.
Schlimm, D. [2006]. "Axiomatics and progress in the light of 20th century philosophy of science and mathematics." *Foundations of the Formal Sciences IV: The History of the Concept of the Formal Sciences*. Ed. by B. Löwe; V. Peckhaus; and T. Räsch. Vol. 3. Studies in Logic. London: College Publications. 233–53.
Stanley, J. [2004]. "On the linguistic basis for contextualism." *Philosophical Studies* **119**. 119–46.
Stich, S.P. [2001]. "Plato's Method Meets Cognitive Science." *Free Inquiry* **21**. 36–8.

Swain, S.; J. Alexander; and J.M. Weinberg [2008]. "The instability of philosophical intuitions: Running hot and cold on Truetemp." *Philosophy and Phenomenological Research* **76** (1). 138–55.

Weinberg, J.M.; S. Nichols; and S. Stich [2001]. "Normativity and Epistemic Intuitions." *Philosophical Topics* **29** (1/2). 429–60.

ESSAY 9

Who Speaks Mathematics?
A Semiotic Case Study

Roy Wagner

This paper uses a case study to perform a semiotic analysis of the enunciative position in a mathematical text. My attempt is to demonstrate the complexity of this enunciative position in order to link the research of mathematical textual practices to perspectives of critical theory.

Traditional philosophy of mathematics accounts for the mathematical subject from epistemological and ontological vistas. The enunciative position is hardly ever taken as a main concern, but it does come up, and requires accounting for. Links between various forms of intuitionism and platonism on the one hand and Kantian or Husserlian transcendental subjectivity on the other appear to provide such accounts (e.g. Atten [2007]; Atten and Kennedy [2003]; and Tieszen [2000]). Formalism, on the other hand, flirted with crossing the line between subjectivity and mechanical processing, especially around the notion of Turing's thinking machine.

The linguistic turn brought with it Wittgenstein's ([1975] and [1978]) philosophy of mathematics, where the subject position is populated by an interactively learning, rule following, symbol manipulating form of life. More recent sociological accounts developed in the context of Science Studies led to the articulation of a social subject of mathematics (for a 'generic' version see Ernest [1998], for some culture and gender specific discussions see Ernest [1994]). The only theoretic structure, however, which focuses primarily on the role of the mathematical text's enunciative position as a semiotic agency, was presented by Rotman in his [1993] and [2000].

What seems to be common to all the above discussions of mathematical subject and enunciative positions is an overwhelming tendency to avoid working with actual mathematical texts. The analyses are justified by imagined, generic, and abstract mathematical texts, with an occasional allusion to some basic arithmetic or formal logic. While this might be somewhat excusable when discussing mathematical transcendental subjects, this is no longer legitimate when discussing social subjects and the semiotic function of a mathematical enunciative position (some recent research into mathematical semiotics does analyse concrete textual examples, e.g. Lefebvre [2000], Netz [2004], and Grosholz [2007], but such research is not centred around analyses of the enunciative position).

In order to rectify this situation, this paper will perform a hands-on analysis of the enunciative position in a specific case study. The texts I chose to investigate are Gödel's proofs of his first incompleteness theorem. Such concrete analysis, of course, has its price: It cannot be immediately generalised to *any* mathematical text. Indeed, Gödel's texts have many exceptional features. But the generalisability of this analysis depends on the articulation of the enunciative position in Gödel's text, which, as the examples quoted in this paper demonstrate, is not all that exceptional. At any rate, regardless of how (a-)typical these texts are, even a single case study can demonstrate the potential complexity of mathematical enunciative positions, and embark on a journey to counter stereotypes that reduce them to simplistic, robust, or clear and distinct arrays. I hope that the restricted *vignette* which follows will motivate analyses of other case studies, which may eventually allow for a comparative semiotic study of the multiform variety of contemporary and historic mathematical enunciative positions.

I read Gödel's proofs in two texts: van Heijenoort's 1967 translation of the original paper from 1931, and the 1965 published notes of the 1934 Princeton lectures. Both versions were approved and revised by Gödel himself. References to the texts will be denoted by G31 and G34 respectively, and page numbers will refer to the Gödel [2003] edition.

I made an effort to keep the analysis below as untechnical as possible, and to choose examples which do not require going into mathematical detail. But some introduction to the argument is, of course, indispensable. One of the best self contained introductions to Gödel's proof for an uninitiated reader is, as far as I can tell, still Nagel and Newman [1958].

Theoretically, the analysis in this paper draws mainly on Barthes' [1974] *S/Z* and Foucault's [1972] *Archeology of Knowledge*. Earlier drafts of the work presented here had more explicit references to the concepts and techniques developed in these books. However, the streamlining of the presentation left little explicit evidence for the role of these authorities in the formation of this work, with the exception of a couple of quotations, some guiding questions, and a general mood.

The main conclusion of my analysis is that the mathematical text's successful functioning (its making sense) depends on an enunciative position that forces an overlap between voices and modalities, which are not quite clearly distinct. My claim is that the text makes sense by combining conflicting voices and modalities, and moreover, would fail to make sense if it were to clearly distinguish and separate these conflicting voices and modalities. This situation is not a failure of Gödel or of mathematics, and is not a hitch that needs mending by philosophers or mathematicians. This situation, I claim, as in the case of many other kinds of texts, is a constitutive condition for the successful functioning of a mathematical text.

One of the referees of this paper made the following objection concerning my level of analysis: "In case my complaint about remaining at the rhetorical level is not clear, let me point out that the standard epsilon–delta way of writing about limits, which has to do with the succession of quantifiers for all epsilon there exists a delta . . . This is sometimes written as a mock dialogue. If you choose epsilon, then I can choose delta . . . That does not turn the argument into such a dialogue. If one were to analyse such a dialogue, one would have to reconstruct it in terms of quantifiers to see the structure of the argument." My stance in this paper is diametrically opposite. I am concerned here with mathematical practice, not with formal reconstruction. Mathematics can be practiced without formal reconstructions, but has never historically existed without surface dialogues. I do not mean to dismiss formal reconstructions, but I do object to the dictum that one "would have to reconstruct it in terms of quantifiers" to analyse the successful functioning of mathematics. I am trying to study how mathematics works, not how it should work.

The analysis is not completely at odds with a Wittgensteinian analysis. However, where a Wittgensteinian analysis seeks to set apart and distinguish the different language games involved in mathematical practices (e.g. Wittgenstein [1978]: 118–22, especially §8), I point out that it is precisely the 'undecidable' meshing together of different language games that allows mathematics to make sense.

My approach employs techniques of structural discourse analysis in order to diagnose the different voices and modalities in the text. However, by problematising the articulation into voices and modalities my analysis undermines the distinction between elements of the analysis *found within the text* and analytic elements forced *on the text from without*. For these reasons the semiotic analysis presented in this paper can be labeled post structural.

The project presented here may be continued in two ways. The first is broadening the scope of the analysis towards an overall post-structural semiotic analysis of the text. This is done in Wagner [2008] (for a different kind of post structural semiotic analysis of mathematical texts see Wagner [2009a]). The other direction is to extend the semiotic analysis of the enunciative position to other mathematical texts. I hope that this project too will get the attention it merits.

The paper is organised as follows. The first section brings up the structural complexity of the personal pronouns 'I' and 'we.' This analysis is carried out in a tradition that can be traced to many thinkers. Here I chose to refer to Benveniste due to his semiotic focus, and due to his concern with the quotability of the personal pronoun. The second section studies the enunciative position in the mathematical texts through the verbs it contracts. The analysis engages with Rotman's division of the mathematical enunciative position into three persons, critically examines the tenability of this division, and offers a more complex picture of the interaction between the various voices included in the enunciative position. The third section goes on to examine the enunciative position through the temporality, spatiality, and modalities that it employs. Here too a modal web (or perhaps *rhizome*) is traced, which makes sense through its incongruent superpositions. The fourth section looks at the articulation of the enunciative position in the texts as polar to a constructed coalition of its contemporaries. The final section complements the above picture with a brief look at the few instances of placing the texts' enunciator in an object position. This last section also serves to conclude the argument and engage in some brief theoretic reflection.

1 We and I

The enunciative position in Gödel's texts is marked by a "we." In order to study the enunciative position as articulated in Gödel's text, we shall analyse the grammar of this "we." But before we do, I feel bound to acknowledge that we needn't read too much into this "we." "We" is simply part of the code. Its use is as imposed upon Gödel as is using "I" when I speak to a friend. So we needn't read too much into "we," no more, say, than Emile Benveniste reads into "I."

"I," Benveniste explains, is an interface between an individual and language. "Language is so organised that it permits each speaker to *appropriate to himself* an entire language by designating himself as *I*." It is this precise interface, according to Benveniste, which turns the individual into a subject, and without which subjectivity could not be formed. "*I*" achieves that as it "refers to the act of individual discourse in which it is pronounced" (Benveniste [1971]: 226). In fact, "*I* is 'the individual who utters the present instance of discourse containing the linguistic instance *I*'" (Benveniste [1971]: 218).

This unified functional view of appropriative "I" would be very convincing, if it weren't for the multitude of different instances of "I" appearing in Benveniste's text. Despite the statement of the first quotation above the instances of "*I*" in the previous paragraph do not function to appropriate language to individuals, but are signifiers referring to this appropriative interface (as witnessed by the italics and the constructions "*I* refers" and "*I* is," rather than 'I refer' and 'I am').

But this is not all. The non actual, non appropriative "*I*" breaks down into smaller constituents. The last quotation above, under pains of circularity, must generate two different grammatical positions ("referent" and "referee," according to Benveniste). It begins with "*I* . . . the individual," and ends with "the linguistic instance *I*." Surely, even if "Man is a sign" (Peirce [1958]: §314), or as Benveniste puts it, "'subjectivity' . . . is only the emergence in the being of a fundamental property of language" (Benveniste [1971]: 224), it is not the individual who is a linguistic instance (at least not according to any point of view that insists on articulating the individual and the subject as distinct).

We so far have an appropriative "I," which a speaker designates himself as in order "to *appropriate to himself* an entire language", the reported, non-actualised "*I*," who is the speaking individual who uses this interface, and the "linguistic instance *I*." If only things were so simple! consider that "the indicators *I* and *you* cannot exist as potentialities, they exist only insofar as they are actualised in the instance of discourse" (Benveniste [1971]: 224). Here "*I*" is not that which a speaker designates himself as, but that which a speaker designates himself with (the speaker Emile does not designate himself as an indicator when he designates himself as "*I*"). Nor is it the uttered linguistic instance, because an indicator is far more functional than a mere linguistic instance. But this "*I*," that is, the indicator, is obviously not the "*I*" who is "'the individual who utters . . . ,'" because indicators are not individual people. We therefore have a fourth, indicator "*I*." But then we're still not done. The "*I*," which appears in "the indicators *I* and *you* cannot exist as potentialities" can't be an indicator either, because indicators have, in that very sentence, been denied the possibility of existing "[un]actualised in the instance of discourse," whereas the indicative function of the last "*I*" is still strictly a potentiality, the very potentiality which would indicate when and if it were eventually actualised in discourse. It is, so to speak, the signifier of the indicator.

It seems that whenever we try to articulate "I," another structural dimension is forced upon it. Whatever instance of "*I*" we are given, science appears to objectify its function, and then signify that object.

But there is yet another, deeper rooted reason for the plurality of "I," which goes beyond the machine of infinite regress. For if "*I*" referred to a constant individual, "a permanent contradiction would be admitted into language, and anarchy into its use." This is especially apparent when considering the quotability of "I", that is the fact that "If I perceive two successive instances of discourse containing *I*, uttered in the same voice, nothing guarantees to me that one of them is not a reported discourse, a quotation in which *I* could be imputed to another." On the other hand, and at the same time, "It is by identifying himself as a unique person pronouncing *I* that each speaker sets himself up in turn as the 'subject'" (Benveniste [1971]: 218, 220, 226).

I am led to conjecture an extension to Benveniste's claim, which the following paragraphs will attempt to validate: *It is not only intersubjectivity, but substantial components of the very experience of making mathematical sense too, which arise, at least in part, from confounding all these instances of "I" and all these functional positions.*

2 What We Do

How does the enunciative position in Gödel's text, "we," fare compared to Benveniste's "I"? Gödel's 1931 text contains only three instances of "I," and the 1934 text only one. These instances of "I" are used in an acknowledgment, a reference to previous work, a reference to a lecture, and to comment on the relation between Gödel's work and Hilbert's programme. But the person narrating, articulating, and conducting the proof is "we." And since analysing the grammatic structure of the personal pronoun in the previous section has led us to a futile proliferation of structural positions, let us concentrate on what "we" do. Taking an inventory of Gödel's texts we find that "we" abbreviate, accomplish, add, adjoin, allow, apply, assign, associate, assume, attach, carry out, come, compare, consider, construct, deduce, define, denote, depend, describe, derive, eliminate, employ, establish, exclude, express, find, find convenient, generalise, give, have, include, insert, intend, let, list, make, make use, map, mean, note, observe, obtain, order, proceed, prove, put, replace, require, restrict, say, see, shift, show, sketch, substitute, take into account, turn to considerations, understand, use, wish, and write.

To make sense of what "we" do, let us follow Foucault, who suggests that a mathematical text articulates various subject positions. "Take the example of a mathematical treatise. In the sentence in the preface in which one explains why this treatise was written . . . the position of the enunciative subject can be occupied only by the author . . . only one possible subject . . . On the other hand, if in the main body of the treatise, one meets a proposition like 'Two quantities equal to a third quantity are equal to each other,' the subject of the statement is the absolutely neutral position, indifferent to time, space, and circumstances, identical in any linguistic system, and in any code of writing or symbolisation, that any individual may occupy when affirming such a proposition. Moreover, sentences like 'We have already shown that . . . ' necessarily involve statements of precise contextual conditions that were not implied by the preceding formulation: The position is then fixed within a domain constituted by a finite group of statements . . . The subject of such a statement . . . will not be described as an individual who has really carried out certain operations, who lives in an unbroken, never forgotten time" (Foucault [1972]: 94).

We may try to make systematic sense of these positions by relating them to the list of verbs above. Different verb types may articulate different functional positions taken by "we." Fortunately, such typological analysis has already been conducted in the general context of mathematical texts by Brian Rotman in his [1993] and in the first chapter of his [2000]. His account, which claims to derive from Peirce, is as follows.

Three characters take part in a mathematical text.

(1) The first is called the *Person*, who speaks in the meta-Code. He speaks about mathematics, but not in mathematics. He considers the idea or story behind the proof. He has access to natural language, the personal pronoun "I," and indexicals such as "here," and "now". His function is to be persuaded by the proof and to understand it.

(2) The second character is the *Subject*. The Subject is the entity that defines, derives, considers, and proves. His pronouncements are, according to Rotman, voiced in a collective imperative: "[L]et us consider, define, demonstrate . . . " (Rotman [1993]: 71). The Subject is abstracted of any temporal, spatial, and cultural con-

siderations. "The subject's psychology, in other words, is transcultural and disembodied" (Rotman [2000]: 15). The Subject makes mathematical statements, which are, in fact, predictions on the outcomes of sign manipulations. Thus, $x + y = y + x$ is the prediction that if he substitute any number-signs for x and for y, and manipulate these signs according to the orders coded in the '+' sign, the results of the two sides of the equality will be the same.

(3) The last character is the *Agent*. The Agent is a reduced image of the Subject, who performs the sign manipulations, of which the Subject makes predictions. He is the one adding, counting, substituting in a sort of thought-experiment or a dream which the subject is experiencing. The "Agent, unlike the Subject, has no ability to imagine and can only respond to signs in their truncated, skeletonised form as signifiers devoid of intentioned meaning. In other words, the Agent is considered as an automaton, a wholly mechanical and formal proxy for the Subject" (Rotman [1993]: 76). The Agent's actions are expressed in an exclusive imperative mood[1]. (Add! Count! Integrate!)

The "Person constructs a narrative, the leading principle of an argument, in the meta-Code; this argument or proof takes the form of a thought experiment in the Code; in following the proof, the Subject imagines his Agent to perform actions and observes the results; and in the light of the narrative, the Person is persuaded that the assertion being proved—which is a prediction about the Subject's sign activities—is to be believed" (Rotman [2000]: 35). "The hierarchy of semiotic agencies here, from imagined Agent to imagining Subject to indexically conscious Person structuring a thought experiment is isomorphic to that on which any dream rests: the Agent maps onto the figure dreamed about, the Subject the dreamer dreaming the dream, and the Person the dreamer awake, consciously interpreting and recognising the dream . . . the dream-code . . . is restricted in various ways, not least by the lack of the ability to recognise the dream *as* dream, which makes it impossible *for the dreamer* to articulate the dreamer's kinship to the imago he or she dreams into being. Likewise, the restrictive nature of

---

[1] An exclusive imperative is an imperative which does not include the person giving the order. 'Eat!' is an exclusive imperative. 'Let's eat!' is an inclusive one.

the Code in which the Subject operates, in particular its lack of indexicality, prevents the mathematical Subject from articulating the status of *its* created fiction" (Rotman [1993]: 78–9).

To explain his division (which is akin to—but does not derive from and is not identical with—Foucault's), Rotman appeals to the imagery of the dream. But his account fails to acknowledge a very unsettling, albeit not too uncommon, occurrence: *lucid dreaming*. Lucid dreaming is the experience where a dreamer is aware that she is dreaming, and where control over the dream is negotiated between various subjective parts. We must not neglect the possibility that the dream-code may be lucid, and so, occasionally, we should consider waking from the lucid dream. Such experience, I recall, can be extremely unpleasant. Prominent physicist Richard Feynman, however, provides a somewhat more optimistic account of this unpleasant experience. "During the time of making observations in my dreams, the process of waking up was a rather fearful one. As you're beginning to wake up there's a moment when you feel rigid and tied down, or underneath many layers of cotton batting. It's hard to explain, but there's a moment when you get the feeling you can't get out; you're not sure you can wake up. So I would have to tell myself—after I was awake—that that's ridiculous. There's no disease I know of where a person falls asleep naturally and can't wake up. You can *always* wake up. And after talking to myself many times like that, I became less and less afraid, and in fact I found the process of waking up rather thrilling—something like a roller coaster: After a while you're not so scared, and you begin to enjoy it a little bit" (Feynman [1985]: 50). Let us, then, muster the courage to elucidate Rotman's "dream," and talk ourselves awake.

Rotman's divisions are based on both semantic and morphological markers. Morphologically, the Person is the one using the pronoun "I" as well as indexicals and cultural markers. The Subject is characterised by collective imperatives, and the Agent by exclusive imperatives. Unfortunately, such markers are absent from Gödel's text. Rotman claims that "Even a cursory examination of an arbitrary chosen item of mathematical communication will reveal two fundamental features of mathematical discourse: its organisation as an exhortatory, command-giving formalism and its complete lack of any indexical terms" (Rotman [1993]: 71). I do not dispute that if one tries hard enough, one could find such mathematical texts (some concise accounts of geometric constructions might be a good place to look). But our "arbitrary" texts, Gödel's 1931 and 1934 texts (and, as far as I can tell, most mathematical texts), do not conform to these characterisations.

First, there are very few imperatives in these texts. Perhaps the imperative mood was considered ill-suited for civilised written communication in the Vienna and Princeton circles. Perhaps it never occurred to writers in these circles to dominate a mathematical text with the imperative mood. Instead of imperatives, we have the frequent use of the indicative mood in both active and passive voices, often attributed to the character "we." In addition, the texts contain 9 "here[s]" and 33 "now[s]," some of which would be difficult to marginalise to the "meta-Code," as they are used to order and demarcate the formal deductive process. These indexicals form part of a network of locative and temporal adverbial structures that operates in the text.

This does not mean that Rotman's analysis has broken down. The lack of morphological markers does not mean we cannot establish a structural division of the enunciative position. We may regroup verbs into subsets, and identify the Subject with occurrences of "we" bundled up with some verbs, the Agent with occurrences of "we" bundled up with others, and the Person as related to yet another set of verbs and adverbials. But here we should take into account one of Barthes' observations concerning the threading of sequences of verbs. "Actions," he explains, "can fall into various sequences which should be indicated merely by listing them, since the [sequence of actions] is never more than the result of an artifice of reading: whoever reads the text amasses certain data under some generic titles for actions ... and this title embodies the sequence; the sequence exists when and because it can be given a name, it unfolds as this process of naming takes place, as a title is sought or confirmed; its basis is therefore more empirical than rational, and it is useless to attempt to force it into a statutory order; its only logic is that of the 'already-done' or 'already-read'" (Barthes [1974]: 19). A formal grouping of verbs risks being an arbitrary empirical compulsion rather than a well grounded structural conclusion.

Bearing this cautionary statement in mind, let's try to allocate some verbs appearing in the texts to the Person, Subject, and Agent, and see whether we can distinguish which verb belongs to which character, or whether this distinction is an interpretive framework imposed on the mathematical text. We consider three examples. I do not introduce the details which are required for a technical understanding of the quotations below. But, as will become evident, the distribution of actions between characters does not depend on such understanding (we only note that *PM* is the initials of Russell and Whiteheads' formal system *Principia Mathematica*).

EXAMPLE I: "[W]e can, for example, find a formula $F(v)$ of *PM* with one free variable v (of the type of a number sequence) such that $F(v)$, interpreted according to the meaning of the terms of *PM*, says: v is a provable formula." Footnote: "It would be very easy (although somewhat cumbersome) to actually write down this formula" (G31: 147).

This text appears in the introduction, in a paragraph that opens with the statement: "Before going into details, we shall first sketch the main idea of the proof," and must therefore be attributed to the Person. But it is obviously not the Person who "find[s]" formulas. The Person is the one who might try to "write down" the formula, since he is the only one who has enough embodiment to experience how "cumbersome" it is. But "find[ing]" the formula, which in the context of this example is a tedious formal procedure, is a task which the Subject should narrate and the Agent should perform. Now, which is the character who does the "interpret[ing]"? Is it the Person who has an overview of all the different layers and significations of the argument, or is it the Subject, for whom interpreting would stand for establishing a correspondence between the provability of a meta-statement and the provability of a formula, a correspondence that the Agent can verify by symbolically manipulating the number sequence substituted for v? And once the Person, Subject, and Agent have thus collaborated, how come it is the formula, rather than any one of them, which, like Balaam's ass, is suddenly conferred with the power to "say"?[2]

---

[2]Some logicians object that formulas don't say, and shouldn't be described as saying. This normative judgment is irrelevant here. I am analysing how a specific mathematical text works. According to this text, formulas do say. In Wagner [2009b] I take more of an issue with attempts to silence formulas.

EXAMPLE II: "[W]e can define relations to be classes of ordered pairs, and ordered pairs to be classes of classes; for example, the ordered pair a, b can be defined to be $((a), (a, b))$" (G31: 153).

The notions "relation," "ordered pair," and "class" are all already determined from the point of view of the mathematical Person. The authority to represent one by another is therefore contingent on the Person's consent, and it is therefore he who "can define." But the one to *actually* make the definition, and perhaps prove the formal adequacy of the definition, is the Subject. Finally, it is the Agent, who, whenever hearing the Subject speak of "the ordered pair a, b" must replace such sign sequence by "$((a), (a, b))$", and only then carry out his further manipulative tasks. In the Person's voice, the above statement reads: I agree to define. In the Subject's voice it reads: I define (and, perhaps, verify that the definition is adequate). In the Agent's voice it reads: I make the symbolic manipulations required by the definition.

EXAMPLE III: "We have noted that $x\mathcal{B}y$ is a recursive relation; and we can also prove that $\sigma(x, y)$ is recursive, where $\sigma(x, y)$ is the number of the formula which results when we replace all free occurrences of w by z_y in the formula whose number is x" (G34: 359–60).

This statement seems to have some verbs relating to the Subject ("note" and "prove"), and some to the Agent ("replace"). However, it also contains temporal and locative elements. The noting is articulated as having occurred in the past. This is a psychological and relative past. If for instance, I skip the text's discussion of recursive relations, and move directly to the main argument (as in fact I first did), then my reading Subject and Gödel's writing Subject get out of synch, and manifest a temporal relativity, whereas the mathematical Subject is supposed to be free of such relativity. Gödel's Subject's present-perfect "have" is my Subject's future-simple "will." This de-synchronisation is in fact explicitly supported by the text, which states that the discussion of recursive functions is "a parenthetic consideration that for the present has nothing to do with the formal system P" (G31: 157). The discrepancy between Gödel's Person and mine has seeped through to create a discrepancy between Gödel's Subject and mine. The embodied Person is allowed to carry the Subject along with it.

On the other hand, the locative preposition "in" is, as far as the text is concerned, an absolute disembodied location, which is invariant to any Subject and Agent (I do not claim that it is an absolute invariant, but only that the text articulates it as such). That this locative be a disembodied invariant is demonstrated by the text's claim that such manipulations can be mechanised. However, even Rotman allows a restrictedly embodied manifestation of locality into the mathematical code, as provided by the *zero* point of reference. "The point to be made here about '0' is that in practice its two senses—Coded number and metaCoded origin—are inextricable from each other: In the course of manipulating the number sign, the meta-sense is always present shadowing it, being part of another layer of meaning which adjoins and penetrates the formal layer available in the code" (Rotman [1993]: 75).

The three examples above *are* analysed according to Rotman's articulation, however problematic and obscure the analysis turned out. There is, however, nothing *in* the text to impose such analysis. Gödel's verbs and pronoun "we" do not seem to distinguish between the Person, Subject, and Agent, but rather to bundle them seamlessly together. We saw that a single verb can order into action all three characters, and that their realms of authority are not clearly distinct. On top of this obscurity, a writer who were to take as a point of departure Foucault's division quoted above, would likely end up with a somewhat different character distribution.

I find no explicit textual evidence that such triple articulation permeates the text. And yet, we cannot ignore the fact that we *can* indeed *rewrite* the text so as to conform to Rotman's articulation. The fact of this capacity must not be dismissed, because it is still a fact *of* the text. It is still a fact of the text that such a reading can be sustained, and as such this fact may very well be relevant to the question of how the text works. This fact can contribute to our understanding of how the text interacts with possible readers, even though it is no more a fact *of* the text than the effect of an *outside* intervention. The fact that a structure, which is not *in* the text, can be imposed *upon* the text, that such imposition does effect some changes, but does not entirely break the text apart, and that such imposition can appear, at least at first sight, quite convincing—these facts testify to a critical feature of mathematical texts: openness[3]. This is the same sort of openness, or rather tolerance, which allows the extraction from (or imposing on) mathematical texts of platonist, logicist, formalist, and intuitionist positions—even though these positions needn't be discoverable from analysing mathematical texts according to terms they explicitly state.

What we impose on the text from the outside is not without precedent inside. The roles suggested by Foucault and Rotman, as well as platonist, logicist, formalist, and intuitionist metaphysics/epistemologies are not completely foreign to the text. They are not arbitrary chance constructs that have nothing to do with the mathematical text. They can all be supported to an extent by evidence excavated from within the text. Indeed, If this had not been the case, no one would have been able to successfully impose these constructs on mathematical texts, or at least not in ways that appear convincing to the extent that they do. The mathematical text communicates with various structures, which are not completely foreign to it, but which are not properly endorsed by it either. This is precisely how a mathematical text can signify: by communicating with its outside, by incorporating inside a residue of its outside, by denying its outside the clear and distinct metaphysical status of outside.

This diagnosis of openness lies well within the range spanned from Barthes' and Foucault's notions of authorship in Barthes [1977] and Foucault [1984], through Eco's [1989] conception of *The Open Work*, to Derrida's maxim that "there is no outside-text" (Derrida [1976]: 158). The novelty is in demonstrating that mathematical texts are not outside this range. Without such openness, without the text's incorporation of externally imposed enunciative positions, the text would simply not communicate with 'foreign' readers and hence soon fail to signify. But imposing upon "we" the three (Person, Subject, Agent) who are one is still awkward and contrived, and it is the interpretive gap between the

---

[3] An entirely different question is *for what purpose* we should want to impose upon a text Rotman's (or any other) articulation. Rotman uses his imposed structure to argue for a radically finitist view, which I respect, and, with some modifications, endorse. Rotman's rearticulation has its merits. But it is not 'discovered' within the text.

text and the reconstruction that I seek to bring up and explore. Facing "we" with the question "What is thy name?" one should not confine to the asylum the possibility of "saying, My name is Legion: for we are many" (Mark 5:9).

3   When, Where, and How We Do It

Here is what we learnt from analysing the enunciative position's actions: that the text tolerates, at a price, divisions which are imposed on it, and allows trace evidence of such divisions to be excavated from within it. This openness to an outside (which undermines a clear inside/outside division, because it blurs the line between imposing something from the outside and discovering it inside) is a crucial constituent of the text's semiotic capacity, but leaves us without a concrete analysis of the text's enunciative position. We shall therefore try to determine this position via the adverbials and modalities interacting with this position.

Temporality makes its mark on the text in various forms. Temporality appears in formal manipulations, which may take place one "after" another, as in: "[W]e must therefore (1) eliminate the abbreviations and (2) add the omitted parentheses" (G31: 156). Temporality also appears along the course of the argument, as in: "We have noted that xBy is a recursive relation" (G34: 359), and in "As will be shown later, however, the converse does not hold" (G31: 173). Finally, temporality intervenes in reviewing and interpreting the text: "[S]uch a proposition involves no faulty circularity, for initially it [only] asserts that a certain well-defined formula . . . is unprovable. Only subsequently (and so to speak by chance) does it turn out that this formula is precisely the one by which the proposition itself was expressed" (G31: 152).

But to appreciate the nature of this temporality, an important clue is provided by the transmutation of 'outside world' temporality into 'mathematical' temporality. This happens in the following passage: "Suppose that on 4 May 1934, A makes the single statement, 'Every statement which A makes on 4 May 1934 is false.' This statement clearly cannot be true. Also it cannot be false, since the only way for it to be false is for A to have made a true statement in the time specified and in that time he made only the single statement" (G34: 362).

This form of the liar paradox is then translated into mathematical form. To do that, Gödel first constructs an explicit enumeration of all formulas in his formal system. Then a formula is constructed claiming that formula number so-and-so has a certain property F (interpreted as being false). The construction is so designed that the number of that formula is the very number so-and-so mentioned inside the formula. The formula therefore says of itself that it has property F (namely, is false). But where in the transition from "4 May 1934" to the self-referential statement did temporality disappear?

Temporality has been converted into enumeration. The date "4 May 1934" served only to designate A's claim. This designation device is transformed into Gödel's enumeration. The form of temporality acknowledged here by the text is so reductive, it is not even properly serial. The numbers conferred upon formulas have nothing to do with the order in which they appear. Time is but a pool of events and designations, which excludes co-occurrence, but which does not display any equivalent of duration (some possible exceptions will be treated below).

But the most interesting aspect is not the articulation of time into a set of mere mutual exclusions, but the ambiguous relation between the enunciative position and this temporality. On the one hand, the enunciative position is within time, as the above quotations explicitly mark. However, the same position has the capacity to review the entire temporal pool. What "will be shown" and what "we have noted" is readily available for the current moment of the discussion. Whatever happened "initially" and "subsequently" in the quotation standing four paragraphs above is bundled together in a contemporary moment where the proposition is self-referential, but "involves no faulty circularity." Even hypothetical moments of an unspecified time can all be bundled up and reviewed together: For any relevant system of axioms κ, "there always are propositions . . . that are undecidable . . . as soon as [some specific proposition] is not κ-PROVABLE" (G31: 195).

The enunciative position, which is within time, and at the same time can observe time and even form it by the operation of enumerating its minimal elements—this complex position contributes to the complex process of semiosis. First, temporality as a set of mutual exclusions registers the distinct elements (formal propositions) studied by the mathematical text. Second, temporality as a sequence of past and future occurrences helps order and narrate the text for the reader. Finally, by taking a stance that can inspect all these different times Gödel counters the grave accusation that his argument is groundless, and shows that his "proposition involves no faulty circularity." In so far as temporality is reduced to the articulation of mutual exclusive elements, there is a sort of circularity (self reference)—the referent formula and referee formula are one and the same. But in so far as temporality has a before and an after, in so far as "initially it [only] asserts" and "[o]nly subsequently (and so to speak by chance) does it turn out"—in that sense we obtain a non-circular turn of events.

The extraction of meaning from the text depends here on instilling a difference across the self-referring instance. The fact that the order of arguing and reading is not absolute, but revisable by any reader, does not weaken the argument that depends on its distinct temporal sequencing. The fact that one has to adopt an omni-temporal point of view in order to complete the story does not weaken it either. But the enunciative position must be able to endorse *all* these conceptions of temporality in order to put Gödel's argument together. The argument's validity depends on a multiplicity within the enunciative position. If the enunciative position were severed into several positions, each endorsing only one conception of time, the argument would simply not stick together[4].

Moving from temporality to spatiality, space in the text is usually reduced to a relative locations in sequences. This relative, disembodied, notion of space is further abstracted by formulations such as: "[A]

---

[4] Some researchers insist on discarding the narrative, and reading the proof 'purely formally,' without retaining the notions of reference and expression operative in the proof. This reading is of course tenable, but leaves the reader in a senseless position. In this reading all that Gödel's theorem provides us with is a couple of highly complex arithmetic propositions concerning some very big numbers. Unless these numbers are allowed to refer to formulas, there is no incompleteness theorem. But we may go even further: If we refuse to let signs refer, all we have is a bunch of symbols that obey some combinatorial restrictions—not even an arithmetical statement. At any rate, I am interested here in following the terms of Gödel's texts, not formalist reconstructions.

formula will be a finite sequence of natural numbers" which is accompanied by the footnote: "That is, a number-theoretic function defined on an initial segment of the natural numbers. (Numbers, of course, cannot be arranged in a spatial order)" (G31: 147). The spatial concept of "sequence" is here reduced to a function providing an ordinal 'location name' for every element in the de-spatialised sequence. The underlying spatial configuration is dismissed and marked impossible due to the non spatial abstraction of numbers.

Spatial notions in the text are rarely related to the enunciative subject. Two of the few exception are: "This makes no difficulty in principle. However, in order not to run into formulas of entirely unmanageable lengths . . . the construction of the undecidable proposition would have to be slightly modified" (G31: 149) and "The proof that [so-and-so holds] is too long to give here" (G34: 359). Here the physical limitations of space, and likely also time (or perhaps spatialised-time), suddenly intervene.

The discrepancy between this last notion of space-time and the ones observed immediately before are better accounted for by an analysis of modality in the text. The text appears to distinguish difficulty "in principle" from "actual" difficulties. Consider, for instance, that "we always understand by 'formula of [the formal system] *PM*' a formula written without abbreviations (that is, without the use of definitions). It is well known that [in *PM*] definitions serve only to abbreviate notations and therefore are dispensable in principle" (G31: 147). It is obvious that abbreviative definitions cannot be dispensed with, if mathematics is something to be practiced by physically and culturally constrained humans of a kind we tend to meet. The above statement, therefore, serves, rather than to make a claim, to articulate the field where the argument is to apply. The argument is to apply in the realm of "in principle."[5] One should bear in mind that in practice mathematicians hardly ever write proofs in as precisely articulated formal systems as "*PM*" at all.

Similar considerations apply to the already quoted statement, which

---

[5]This realm opens up a discussion of the multitude of conceptual layers operative in the proof, especially around different notions of the term meaning. This analysis is conducted in Wagner [2009*b*].

appears on the same page: "It would be very easy (although somewhat cumbersome) to actually write down this formula." Easy, perhaps—but not humanly feasible. Today it may be feasible to program a computer to output such a formula, but extremely cruel, and quite likely unfeasible, to expect a human to examine the output. Either way, this option was not available to Gödel. Again, we have a statement that operates in the language game of the "in principle" grammar, where things can be done that cannot *actually* be done.

We may appear to have found a textual support for the Agent vs. Subject–Person division. It is the disembodied, unlimitedly industrious Agent, who can do whatever can be done "in principle." The Person, perhaps even the Subject, are excluded from this capacity. But whereas we do have two distinct modalities in the text, there is no indication that the enunciative position is in fact divided according to these modalities[6]. Consider for instance the statement (also on the same page) "[I]t can be shown that the notions 'formula,' 'proof array,' and 'provable formula' can be defined in the system *PM*; that is, we can, for example, find a formula $F(v)$ of *PM* . . . such that $F(v)$ . . . says: v is a provable formula." The first "can" will in fact be achieved in the text. But the next "can" is precisely the kind that is only achieved "in principle", and the last "can" falls somewhere in the middle, depending on what aspect of "find[ing]" is at stake. The enunciator must endorse both modalities if it is to attain both human accessibility (actual capacity) and the mathematical ideal ("in principle" capacity). The text does not show any sign that the enunciator relates differently to these seamlessly interwoven layers of capacity—and yet these layers are distinguished by the text itself, when it briefly insists on the "in principle" reservation.

What I have been trying to show is that the texts make sense and lend themselves to the extraction of meanings by, on the one hand, supporting, to a certain extent, divisions of the speaker, while, on the

---

[6]Claiming that any grammatical division (or even that any modal division) necessarily entails a division of the enunciative position leads to grotesque results. The imperative and indicative moods are distinct, but they do not necessarily require distinct enunciative positions to function properly. Indeed, Austin's *How to Do Things with Words* ends up questioning the distinctions between indicatives and performatives.

other hand, refusing to be pinned down to such divisions. The enunciative position has been shown to be distributed across various roles—but without actually grounding any stable textual articulation of such roles; it has been shown to endorse various incompatible modalities of temporality, spatiality and capacity—and then bind them together into intelligibility by avoiding clear distinctions followed by explicit syntheses.

The text relies on many diverse positions to explain itself, to make the reader understand; but the text cannot commit itself to any single position, or set these positions as entirely apart, as by doing that it would simply fall apart. The text manages to produce its meaning by distributing differences across the repeated enunciative stance. The semiotic machine embodied by the morpheme "we" is, like Benveniste's "I," a blurring interface that ties a plurality of readers, writers, and their forms of being to texts in ways that make sense. Regardless of popular belief, mathematical texts, such as these, do not make sense by pure formal clarity. They make sense by having it both ways (and more).

4 Who Do We Do It With? (A Slight Detour)

Whoever "we" are, they can't stand alone. "I use *I* only when I am speaking to someone who will be a *you* in my address. It is this condition of dialogue that is constitutive of *person*, for it implies that reciprocally *I* becomes *you* in the address of the one who in his turn designates himself as *I* . . . This polarity of persons is the fundamental condition in language, of which the process of communication, in which we share, is only a pragmatic consequence" (Benveniste [1971]: 224–5).

A polar position to the texts' enunciative position is established in the very first lines of the introduction to the 1931 text. "The development of mathematics toward greater precision," Gödel explains, "has led, as is well known, to the formalisation of large tracts of it, so that one can prove any theorem using nothing but a few mechanical rules. The most comprehensive formal systems that have been set up hitherto are the system *Principia mathematica (PM)* on the one hand and the Zermelo–Fraenkel axiom system of set theory on the other . . . These two systems are so comprehensive that in them all methods of proof are formalised, that is reduced to a few axioms and rules of inference. One might therefore conjecture that these axioms and rules of inference are sufficient to decide *any* mathematical question that can at all be formally expressed in these systems. It will be shown below that this is not the case, that on the contrary there are in the two systems mentioned relatively simple problems in the theory of integers that cannot be decided on the basis of the axioms" (G31: 145).

The footnotes to the second sentence acknowledge Whitehead and Russell, Fraenkel, von Neumann, and Hilbert and Bernays as authors of formal systems. These various authors are invoked as if they belong to a single unified front, binding together the fundamentally logicist agenda of Russell, Hilbert's finitism, Zermelo's realist point of view, and von-Neumann's pragmatist formalism, subjugating all to an ideological field that is negatively marked by the adjective "mechanical." This adjective, along with the verb "reduce," suggests a non-voluntary, non-subjective, and impoverished mathematical practice. Indeed, at the time of writing the paper machines had already had the kind of connotation to be canonically presented in Chaplin's 1936 *Modern Times*. At the same time, the use of the term "mechanical" in the context of formalism is far from obvious. A mechanical approach will only be established a few years later by Church and Turing.

In the above quote a position of 'they' has been set-up and characterised. 'They' are ascribed a mighty achievement. Due to 'their' work, "one can prove any theorem using nothing but a few mechanical rules." An unsuspecting reader may be lured to believe that a certain saturation has been achieved within "large tracts" of mathematics, whereby given a theorem, a proof can (mechanically!) be produced. And indeed, those readers who swallow the bait[7] "might therefore conjecture that these axioms and rules of inference are sufficient to decide *any* mathematical question that can at all be formally expressed in these systems" (G31: 145).

---

[7] The translation appears to be more misleading than the original, which speaks of *Beweisen* (proofs) rather than theorems. But even the original claim is stronger than what was considered as commonly accepted at the time, and indeed, Gödel himself casts a doubt on this claim at the end of the paper (G31: 195).

Note that the constative assertion is reduced by the hypothetical modal "might." The text, however, does not yet clarify to the possibly misled reader that the formalist project allows to transcribe and verify by "few mechanical rules" only those proofs that are *already discovered*. Unproven and unrefuted theorems remain unproven and unrefuted regardless of the formalist project. The modal shift from constative to hypothetical serves to muddle the reader, who is now watching the conjecture he was manipulated into forming subjected to the doubt inscribed in a "might." And indeed, "It will be shown below that this is not the case, that on the contrary there are in the two systems mentioned relatively simple problems in the theory of integers that cannot be decided on the basis of the axioms" (G31: 145). But is it an oversight that an explicit definition of undecidability (there exists a statement such that neither it nor its negation can be proved) is postponed to the bottom of the next page? The specialist may have known exactly which undecidability Gödel was talking about. Other mathematicians were probably still kept in suspense.

The rhetorical structure we have just reviewed is standard. A picture is painted, then questioned, and finally announced invalid. The enunciative position in the text takes advantage of this manoeuvre. For the "It will be shown" to emerge both as a legitimate offspring and a parricidal revolutionary with respect to a certain lineage, a paternal position has to be, in the space of a few lines, both established and denounced. This position is established by forcing together Russell, Hilbert, Fraenkel, and von Neumann into a straw-coalition, ignoring the fact that formalism was extremely young[8], highly unstable, and far from resting on a consensus among the contributors named above. This paternal position immediately falls victim to a parricide performed by branding this straw-formalism with the suspect mark of mechanism, and assigning to it the ambiguous and misleading formulation "axioms and rules of inference are sufficient to decide *any* mathematical question," which formalists would not have dared to claim as achieved.

The positioning accomplished here relies on the ability of a standard rhetorical manoeuvre to impose itself on a discursive field that is still

---

[8] Merely 13 years passed from Hilbert's 1918 formulation; see Kleene's introduction to Gödel's 1931 paper in Gödel [2003]: 126.

underdetermined and emerging. This move provides the rhetorical support (which is, of course, not the only support) for putting it, later in the text, that "The solution suggested by Whitehead and Russell, that a proposition cannot say something about itself, is too drastic" (G34: 362).

Indeed, the tension between belonging to, and breaking away from 'them,' which gives rise to the text's impersonal "it" and collective "we" enunciative positions, is particularly manifest in the manoeuvres around the "proposition that says about itself that it is not provable [in *PM*]," which, as a footnote we have already quoted explains, "involves no faulty circularity, for initially it [only] asserts that a certain well-defined formula . . . is unprovable. Only subsequently (and so to speak by chance) does it turn out that this formula is precisely the one by which the proposition itself was expressed" (G31: 151). The enunciative position endorses here at once the aversion from and attraction to self-reference. No one would deny that the self-referential proposition was constructed with the explicit intention of emulating self-reference inside a formal system. And yet the mere possibility of telling a revisionist history, obviously false *as a history*, relating the accidental genesis of this self-reference, is sufficient to allow contemporary logic both to embrace Gödel's form of self reference,[9] and to be revolutionised by it. The text remains formalist. But into this "mechanic" realm of formalism quietly sneaks an element of "so to speak . . . chance."

But all this rhetorical mechanics and revisionist historical articulation of meta-mathematical positions relates to the preamble to the introduction, and, for fear of being accused of lurking in the margins, we must move back into the core of the mathematical text.

5 Who Does It to Us?

A formalist 'they' position was set-up in order to simultaneously counter and produce the text's enunciative "we" position. But this 'they' position is far from exhausting the stance polar to the enunciative "we." In

---

[9]There are, of course, logicians who altogether deny that Gödel's proposition is self-referential. But I don't believe anyone would deny that self-reference was a "leading principle" in the construction of the proposition.

order to complete the articulation of this polarity, let us track down the few occasions where "we" is demoted from the subject position to the object position "us."

First, we must acknowledge how rare this move is. In both texts there are only nine occurrences of "us," of which three appear in the inclusive imperative "let us." Nevertheless, these occurrences are revealing of the presence of the enunciative position in the much more frequent impersonal and passive constructions[10].

We are told that "if a formal decision ... of the SENTENTIAL FORMULA 17Gen$r$ ... is presented to us, we can actually give ... a PROOF of Neg(17Gen$r$)" (G31: 177). Ignoring the meaning of 17Gen$r$, we observe that polar to "we" a position is established, which may present "us" with mathematical offerings. Confronting this position, "we" becomes a reactive ("giv[ing]" once "present[ed]"), rather than an initiating instance.

But this polar position is not confined to a "giv[ing]" role. "We" can "have before us a proposition that says about itself that it is not provable" (G31: 149–51)—the polar position can speak. In fact, more than just speaking, it can modify the capacities of the enunciative position. A certain proof, for example, "allows us to actually derive a contradiction ... once a PROOF of w ... is given" (G31: 195) (without getting into what it is, such a "PROOF of w" is supposed not to exist).

"We" is therefore a subject position in both senses of the word: subject of and subject to. However, the "proposition that says," which we "have before us," is explicitly produced by "we" in the text. Even the hypothetical object "PROOF of w" above, if provided at all, must necessarily be provided by someone who can present proofs, and therefore share the enunciative position of "we" . "We," it seems, plays a game of catch with itself, or rather, as is now obvious, with themselves.

"We" have many voices, many positions from which "we" speak.

---

[10]I should have, perhaps, analysed the impersonal and passive constructions in the text. Indeed, Foucault writes that the subjective formation "is situated at the level of 'it is said'" (Foucault [1972]: 122). However, since the relation between some of these constructions and the enunciative position is debatable (for instance, in statements of the form 'it follows that . . . '), I preferred to focus on the textual occasions where the voice emanating from the enunciative position is explicitly marked as such (lest I be accused of hearing voices).

"We" speak to and of ourselves and our creations, who in turn, as the above quotations demonstrate, do not hesitate to speak back and challenge. In some sense (a sense I wish to impose) "we" are none but Gödel and I, the reader, in the most intimate moment of reading—intimate but not private, because I know many others have done it with him too, and because, to a certain extent, here I am doing it with him in public. Perhaps these so many discursive partners are the cause of this text being infested with so much meaning. Having intercourse with so many codes (encoded and decoded by so many partners), it is bound to say many things. "We," which is supposed to establish a single common denominator, ends up, it seems to me, forcing the text open to many different entangled codes and complicit partners. So many different entangling codes indeed, that it is no longer clear that one could speak of *any* individual code at all.

But that is a matter for separate discussion. I would like to conclude this discussion with the words by which Barthes described the ambiguous enunciative position in Balzac's short story *Sarrasine*. "Who is speaking?" he asks. "Is it a scientific voice . . . Is it a phenomenalist voice naming what he sees[?] . . . Here it is impossible to attribute an origin, a point of view, to the statement. Now, this impossibility is one of the ways in which the plural nature of a text can be appreciated . . . it may happen that in the classic text, always haunted by the appropriation of speech, the voice gets lost, as though it had leaked out through a hole in the discourse. The best way to conceive the classical plural is then to listen to the text as an iridescent exchange carried on by multiple voices, on different wavelengths and subject from time to time to a sudden *dissolve*, leaving a gap which enables the utterance to shift from one point of view to another, without warning: The writing is set up across this tonal instability (which in the modern text becomes atonality), which makes it a glistening texture of ephemeral origins" (Barthes [1974]: 41–2).

I could use Foucault's words to conclude this section. "In the proposed analysis, instead of referring back to *the* synthesis or *the* unifying function of *a* subject, the various enunciative modalities manifest his dispersion. To the various statuses, the various sites, the various positions that he can occupy or be given when making a discourse. To the discontinuity of the planes from which he speaks. And if these planes are linked by a system of relations, this system is not established by the synthetic activity of a consciousness identical with itself, dumb and anterior to all speech, but by the specificity of a discursive practice. I shall abandon any attempt, therefore, to see discourse as a phenomenon of expression—the verbal translation of a previously established synthesis; instead, I shall look for a field of regularity for various positions of subjectivity. Thus conceived, discourse is not the majestic unfolding manifestation of a thinking, knowing, speaking subject, but, on the contrary, a totality, in which the dispersion of the subject and his discontinuity with himself may be determined" (Foucault [1972]: 54–5).

I would like to endorse this quote, but I am not sure that the terms "regularity," "totality," and "may be determined" are pronounced here with the cautious and critical tone in which I would like to have heard them.[11] I am worried that this quote underplays the capacity of a textual "we" to disperse and integrate, regularise and breach, totalise and re-open, determine and render undecidable. If we do not underplay these capacities, we can perhaps acknowledge not only the rigorous aspect of mathematics, but also its disruptive creative force. And so I shall not be satisfied before I come again (now in a voice drenched with the fluid complexity of so many enunciative distributions compacted into a brief contractive exhale: "We"—no longer opposing a trinity, but embracing its charge, embracing it *too*) to exclaim "My name is Lesion: for we are many."

Acknowledgments

This paper is based on part of the author's Ph.D. dissertation written in Tel Aviv University under the supervision of Prof. Aant Biletzki and Prof. Adi Ophir. I would like to thank my advisors and Prof. Sabetai Unguru for their comments on this work. The dissertation is to appear in book from as Wagner [2009*b*].

---

[11]They most likely are so pronounced in other parts of *The Archeology*.

Bibliography

Atten, M. van [2007]. *Brouwer meets Husserl.* Dordrecht: Springer.
Atten, M. van and J. Kennedy [2003]. "On the philosophical development of Kurt Gödel." *Bulletin of Symbolic Logic* **9** (4). 425–76.
Barthes, R. [1974]. *S/Z.* Trans. by R. Miller. New York: Hill and Wang.
— [1977]. *Image Music Text.* Ed. by S. Heath. London: Fontana.
Benveniste, E. [1971]. *Problems in General Linguistics.* Trans. by M.E. Meek. Coral Gable: University of Miami Press.
Derrida, J. [1976]. *Of Grammatology.* Trans. by G.C. Spivak. Baltimore: Johns Hopkins University Press.
Eco, U. [1989]. *The Open Work.* Trans. by A. Cancogni. Harvard: Harvard University Press.
Ernest, P., ed. [1994]. *Mathematics, Education, and Philosophy: an International Perspective.* London: Falmer Press.
— [1998]. *Social Constuctivism as a Philosophy of Mathematics.* Albany: State University of New York Press.
Feynman, R.P. [1985]. *"Surely You're Joking, Mr. Feynman!"* New York: W.W. Norton.
Foucault, M. [1972]. *The Archaeology of Knowledge.* Trans. by A.M.S. Smith. New York: Pantheon Books.
— [1984]. *The Foucault Reader.* Ed. by P. Rabinow. Hamondsworth: Penguin.
Gödel, K. [2003]. *Collected Works.* 1986–2003. Ed. by S. Fereferman et al. Vol. I. New York: Oxford University Press.
Grosholz, E.R. [2007]. *Representation and Productive Ambiguity in Mathematics and the Sciences.* Oxford: Oxford University Press.
Lefebvre, M. [2000]. "Construction et déconstruction des diagrammes de Dynkin." *Actes de la Recherche en Sciences Sociales* **141–2**. 121–4.
Nagel, E. and J.R. Newman [1958]. *Gödel's Proof.* New York: New York University Press.
Netz, R. [2004]. *The Transformation of Mathematics in the Early Mediterranean World.* Cambridge: Cambridge University Press.
Peirce, C.S. [1958]. *Collected Papers.* 1931–58. Ed. by C. Hartshorne and P. Weiss. Vol. 5. Cambridge: Harvard University Press.
Rotman, B. [1993]. *Ad Infinitum—the Ghost in Turing's Machine: an Essay in Corporeal Semiotics.* Stanford: Stanford University Press.

— [2000]. *Mathematics as Sign: Writing, Imagining, Counting*. Stanford: Stanford University Press.
Tieszen, R. [2000]. "The philosophical background of Weyl's mathematical constructivism." *Philosophia Mathematica* **3** (8). 274–301.
Wagner, R. [2008]. "Post structural readings of a logico-mathematical text." *Perspectives on Science* **16** (2). 196–230.
— [2009a]. "Mathematical marriages: intercourse between mathematics and semiotic choice." *Social Studies of Science* **32** (9). 289–309.
— [2009b]. *Post Structural Readings of Gödel's Proof*. Milano: Polimetrica.
Wittgenstein, L. [1975]. *Lectures on the Foundations of Mathematics*. Ed. by C. Diamond. Chicago: University of Chicago Press.
— [1978]. *Remarks on the Foundations of Mathematics*. Trans. by G.E.M. Anscombe. Revised edition. Cambridge: MIT Press.

ESSAY 10

Diagrammatic Reasoning in Euclid's *Elements*

Danielle Macbeth

Although the science of mathematics has not undergone the sorts of revolutionary changes that can be found in the course of the history of the natural sciences, the practice of mathematics has nevertheless changed considerably over the two and a half millennia of its history. The paradigm of ancient mathematical practice is the Euclidean demonstration, a practice characterized by the involvement of both text and diagram. Early modern mathematical practice, begun in the seventeenth century, is instead computational and symbolic; it constitutively involves the formula language of arithmetic and elementary algebra that one is taught as a schoolchild even today (see Macbeth [2004] and Lachterman [1989]). Over the course of the nineteenth century, this practice gave way, finally, to a more conceptual approach, to reasoning from concepts, for instance, from the concept of continuity in analysis or from that of a group in abstract algebra (see Stein [1988]). This most recent mathematical practice has naturally brought in its train—at least officially, if not in the everyday practice of the working mathematician—a demand for rigorous gap-free proofs on the basis of antecedently specified axioms and definitions. It has also suggested to many that Euclidean mathematical practice is hopelessly flawed.

But Euclidean geometry is not flawed. Although it has its limitations—not everything one might want to do in mathematics can be done in the manner of Euclid—this geometry has, over the course of its two and a half thousand year history, proved to be an extremely successful, robust, and sound mathematical practice, albeit one that is quite different from current mathematical practice. My aim is to clarify the nature of this practice in hopes that it might ultimately teach us something about the nature of mathematical practice generally. Perhaps if we better understand the first (and for almost the whole of the long history of the science of mathematics the only) systematic and fruitful mathematical practice, we will be better placed to understand later developments.

Euclid's *Elements* is often described as an axiomatic system in which theorems are proven and problems constructed though a chain of diagram-based reasoning about an instance of the relevant geometrical figure. It will be argued here that this characterization is mistaken along three dimensions. First, the *Elements* is not best thought of as an axiomatic system but is more like a system of natural deduction; its Common Notions, Postulates, and Definitions function not as premises from which to reason but instead as rules or principles according to which to reason. Secondly, demonstrations in Euclid do not involve reasoning about instances of geometrical figures, particular lines, triangles, and so on; the demonstration is instead general throughout. The chain of reasoning, finally, is not merely diagram-based, its moves, at least some of them, licensed or justified by manifest features of the diagram. It is instead diagrammatic; one reasons *in* the diagram in Euclid, or so it will be argued.

1 Axiomatization or System of Natural Deduction?

In an axiomatic system, a list of axioms is provided (perhaps along with an explicitly stated rule or rules of inference) on the basis of which to deduce theorems. Axioms are judgments furnishing premises for inferences. In a natural deduction system one is provided not with axioms but instead with a variety of rules of inference governing the sorts of inferential moves from premises to conclusions that are legitimate in the system. In the case of natural deduction, one must furnish the premises oneself; the rules only tell you how to go on. The question, then, whether Euclid's system is an axiomatic system or not is a question about how the definitions, postulates, and common notions that are laid out in advance of Euclid's demonstrations actually function, whether as *premises* or as *rules* of construction and inference. Do they function to provide starting points for reasoning, as has been traditionally assumed?[1] Or do they instead govern one's passage, in the construction, from one diagram to another, and in one's reasoning, from one judgment to another? Inspection of the *Elements* suggests the latter. In Euclid's demonstrations, the definitions, common notions, and postulates are not treated as premises; instead they function, albeit only implicitly, as rules constraining what may be drawn in a diagram and what may be inferred given that something is true.[2] They provide the rules of the game, not its opening positions.

Consider, for example, the first three postulates. They govern what can be drawn in the course of constructing a diagram: (i) If you have two points then a line (and only one) may be produced with the two points as endpoints; (ii) a finite line may be continued; and (iii) if you have a point and a line segment or distance, then a circle may be produced with that center and distance. In each case, one's starting point, points, and

---

[1] The underlying assumption perhaps is that, as both Plato and Aristotle thought, any science, including mathematics is, or should strive to be, axiomatic. Insofar as Euclid's system is a paradigm of science, then, it must be axiomatic.

[2] As we will see, Euclid in fact almost never invokes his definitions, postulates, and common notions in the course of a demonstration. They are nevertheless readily identifiable as warranting the moves that are made.

lines, must be supplied from elsewhere in order for the postulate to be applied. And nothing can be done, at least at first, that is not allowed by one of these postulates. But once they have been demonstrated, various other rules of construction can be used as well. For instance, once it has been shown, using circles, lines, and points, that an equilateral triangle can be constructed on a given finite straight line (proposition I.1), one may in subsequent constructions immediately draw an equilateral triangle, without any intermediate steps or constructions, provided that one has the appropriate line segment. Propositions such as I.1 that solve construction problems function in Euclid's practice as derived rules of construction. Once they have been demonstrated, they can be used in the construction of diagrams just as the postulates themselves.

Euclid's common notions, and again most obviously the first three, again govern moves one can make in the course of a demonstration, in this case in the course of reasoning. They govern what may be inferred: (i) If two things are both equal to a third then it can be inferred that they are equal to one another; (ii) if equals are added to equals then it follows that the wholes are equal; and (iii) if equals be subtracted from equals, then the remainders are equal.[3] These common notions manifestly have the form of generalized conditionals, which is just the form rules of inference must take when they are stated explicitly.[4] Furthermore, in this case as well, theorems, once demonstrated, can function in subsequent demonstrations as derived rules of inference. Once it has been established that, say, the Pythagorean theorem is true (I.47), one may henceforth infer directly from something's being a right triangle that the square on the hypotenuse is equal to the sum of the squares on the sides containing the right angle. Indeed, Euclid's *Elements* is so called because the totality of its theorems and constructions provide in this way the elements, rules, for more advanced mathematical work.

Definitions can also license inferences, though, as we will see, they have other roles to play as well. If, for example, a diagram in a proposition contains a circle then the definition of a circle licenses the passage to the claim that its radii are equal. If it contains a trilateral figure,

---

[3] These are, of course, not formally valid rules of inference; they are instead what Sellars has taught us to call materially valid rules.

[4] That rules of inference are inherently conditional in form and essentially general is argued by Ryle ([1950]).

that is, a figure bounded by three straight lines, all of whose sides are equal then the definition of equilateral triangle licenses one to conclude that the figure is an equilateral triangle.

But not everything that happens in the course of a demonstration is governed by an explicitly stated rule, whether primitive or derived. There are two sorts of cases. First, Euclid draws (explicitly or implicitly) various obviously valid inferences, such as that two things are not equal given that one is larger than the other, despite the fact that the rule governing the passage is nowhere explicitly stated. Furthermore, in order to follow a demonstration in Euclid, one must read various things off the relevant diagrams, again according to rules that seem nowhere to be stated. For example, given two lines that cross, or cut one another, in a diagram, Euclid assumes that there is a point at their intersection.[5] The point of intersection seems simply to "pop up" in the diagram as drawn, and is henceforth available to one in the course of one's reasoning.[6] This happens not once but twice in the very first proposition of Book I of the *Elements*, to construct on a given finite straight line an equilateral triangle.

The demonstration begins with the setting out: Let AB be the given straight line. A statement of what is to be done follows: to construct an equilateral triangle on AB. Then the construction is given:

(C1) With center A and distance AB let the circle BCD be described. (This is licensed by the third postulate, though Euclid does not mention this.)

(C2) With center B and distance BA let the circle ACE be described. (Again, the warrant for this, the third postulate, is not mentioned.)

(C3) From point C, in which the circles cut one another, to the points A, B let the straight lines CA, CB be joined. (This is implicitly

---

[5]Thomas L. Heath ([1956]: vol. 1, 242) takes this to be an objection: "Euclid has no right to assume, without premising some postulate, that the two circles *will* meet in a point C." The objection is, Heath says, "a commonplace."

[6]I borrow this use of the expression "pop up" from Kenneth Manders, to whom I am indebted for helping me to appreciate just how important this feature of Euclidean diagrams is. See his [1996] and [2008].

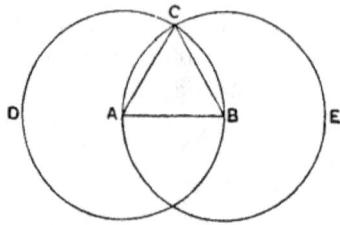

Figure 1

licensed by the first postulate on the assumption that there is such a point C.)

Figure 1 contains the resulting diagram.[7] And the *apodeixis* then follows:[8]

(A1) Given that A is the center of circle CDB, AC is equal to AB. (This is licensed, without mention, by the definition of a circle.)

(A2) Given that B is the center of circle CAE, BC is equal to BA. (This again is implicitly licensed by the definition of a circle.)

(A3) Given that AC equals AB and BC equals BA, we can infer that AC equals BC because what are equal to the same are equal to each other (that is, Common Notion 1).

(A4) Given that AB, BC, and AC are equal to one another, the triangle ABC is equilateral. (This is warranted by the definition of equilateral triangle, *on the assumption that there is such a triangle ABC.*)

This triangle was constructed on the given finite straight line AB as required, and so we are done. In the course of this demonstration, first a point pops up at the intersection of the two drawn circles, and then later a triangle pops up, formed from the radii of the two circles. This

---

[7] This image is taken from the *Elements*: vol. 1, 241.
[8] The word '*apodeixis*' is generally translated as 'proof.' For reasons that will emerge, I leave it untranslated.

sort of thing is, furthermore, ubiquitous in ancient Greek geometrical practice. One simply reads the relevant geometrical objects off the diagrams, apparently without any explicitly stated warrant for doing so. Although nothing can be put into a diagram that is not licensed by one of the given postulates or a previous construction, there seem to be no stated rules governing what, in the way of pop-up objects, can be taken out of it.[9] We will have occasion to come back to this.

2 Generality in Euclid's Demonstrations

I have suggested that Euclid's *Elements* is not best thought of as an axiomatic system (even one that is incomplete, not fully rigorous) because it seems instead to function more like a system of natural deduction. The Common Notions, Postulates, and Definitions function in the *Elements* not as judgments that will, for the purposes of proof, be taken to be true but instead as rules of passage governing allowable moves in the demonstration. We need now to consider whether the diagram provides an instance about which to reason or instead something more general.

We have seen that Euclid's demonstration that an equilateral triangle can be drawn on a given finite straight line begins with a "setting out" (*ekthesis*): Let AB be the given straight line. And this is generally true of Euclid's demonstrations. Although what is to be demonstrated is something wholly general, the demonstration invariably proceeds by way of such a setting out. To anyone familiar with proofs in standard quantificational logic, it is very easy to take this setting out as an analogue of Universal Instantiation. We read, for instance, in Russell's [1956]:

> Given a statement containing a variable x, say '$x = x$,' we may affirm that this holds in all instances, or we may affirm any one of the instances without deciding as to which instance we are affirming. The distinction is roughly the same as that between the general and particular enunciation in Euclid. The general enunciation tells us some-

---

[9] Could Euclid's definitions serve this purpose? Can it be inferred, for example, from the fact that a line is a breadless length that at the cut of two lines there is a point (i.e., something that has no parts)? Perhaps, but more would need to be said.

thing about (say) all triangles, while the particular enunciation takes one triangle and asserts the same thing of this one triangle. But the triangle taken is any triangle, not some one special triangle; and thus although, throughout the proof, only one triangle is dealt with the proof retains its generality.   (Russell [1956]: 64)

In quantificational logic, in order to prove that (say) all A is C given that all A is B and that all B is C, one must first turn to an instantiation of the premises in order that the rules of the propositional calculus may be applied. One reasons, in effect, about a particular case, and then at the end of the proof one takes what has been shown to apply generally on the grounds that no inference was drawn in the course of the proof that could not have been drawn were any other instance to have been considered instead. According to Russell, demonstrations in Euclid work the same way.[10] Indeed, one might be puzzled as to what the alternative might be.

Manders offers a suggestion following Leibniz, namely, that "[D]iagrams [are] *textual* components of a traditional geometrical text or argument, rather than semantic counterparts" (Manders [1996]: 391). That is, the diagram does not provide an instance (a semantic counterpart), but instead what Leibniz calls a "character": "For the circle described on paper is not a true circle and need not be; it is enough that we take it for a circle."[11] But what exactly might this mean? Grice's analysis of the distinction between (as he puts it) natural and non-natural meaning provides the tools for at least an outline of a plausible answer (Grice [1957]).[12]

As Grice points out, there is an intuitively clear distinction between, for example, the sense in which a certain sort of spot on one's skin can mean measles and the sense in which three rings on the bell of a bus can

---

[10] Netz ([1999]: §6) takes essentially the same view: What is to be demonstrated is general but the demonstration itself (that is, the setting out, construction, and *apodeixis*) is particular; its generalizability is "a derivative of [its] repeatability" (ibid.: 246; see also ibid.: 262).

[11] Leibniz [1969]: 84; quoted in Manders [1996]: 291.

[12] See also Dipert [1996].

mean that the bus is full.[13] The task is to provide an analysis of the difference between the two senses of 'means,' natural and non-natural, respectively; and one of Grice's examples to that end is particularly revealing for our purposes. Grice asks us to compare the following two cases:

(1) I show Mr. X a photograph of Mr. Y displaying undue familiarity with Mrs. X.

(2) I draw a picture of Mr. Y behaving in this manner and show it to Mr. X.

And, as he immediately goes on to remark,

> I find that I want to deny that in (1) the photograph (or my showing it to Mr. X) meant$_{NN}$ [that is, non-naturally] anything at all; while I want to assert that in (2) the picture (or my drawing and showing it) meant$_{NN}$ something (that Mr. Y had been unduly familiar), or at least that I had meant$_{NN}$ by it that Mr. Y had been unduly familiar. What is the difference between the two cases? Surely that in case (1) Mr. X's recognition of my intention to make him believe that there is something between Mr. Y and Mrs. X is (more or less) irrelevant to the production of this effect by the photograph . . . But it will make a difference to the effect of my picture on Mr. X whether or not he takes me to be intending to inform him (make him believe something) about Mrs. X, and not to be just doodling or trying to produce a work of art. (Grice [1957]: 282–3)

Although the photograph can serve to convey to someone the fact that Mr. Y is unduly familiar with Mrs. X independent of anyone's intending that it so serve, the drawing cannot. To take the drawing as conveying a message about Mr. Y's behavior, rather than as a mere doodle or as a work of art, essentially involves taking it that someone produced it with the intention of conveying such a message, and that that person

---

[13] In the opening paragraphs of his essay, Grice lists five different ways the use of 'means' differs in the two cases.

did so with the intention that that intention be recognized (and play a role in the communicative act). Only in that case does the drawing mean (non-naturally) that Mr. Y is unduly familiar with Mrs. X.

A photograph has natural meaning in virtue of a (causally induced) resemblance between the image in the photograph and that of which it is a photograph. A drawing, Grice argues, can in certain circumstances have instead non-natural meaning. That is, it can have the meaning or content that it does in virtue of one's intending that it have that meaning or content (and intending that that intention be recognized and play a certain role in the communicative act). So, we can ask, does a drawn figure in Euclid have Gricean natural meaning or instead Gricean non-natural meaning? If it is a drawing of an instance, a particular geometrical figure, then it has natural meaning. It is in that case a semantic counterpart; it is the thing, say, the particular triangle ABC, that is referred to in the text of a demonstration when one judges of triangle ABC that it is thus and so. But perhaps the drawing, like Grice's drawing, instead has non-natural meaning. Perhaps it is not (say) a circle but instead, as Leibniz says, is taken for one. If so, then the Euclidean diagram can mean or signify some particular sort of geometrical entity only in virtue of someone's intending that it do so and intending that that intention be recognized. One's intention in making the drawing—an intention that can be seen to be expressed in the setting out and throughout the course of the construction—is, in that case, indispensable to the diagram's playing the role it is to play in a Euclidean demonstration.[14]

A Euclidean diagram can be interpreted either as having natural or as having non-natural meaning in Grice's sense. If it has natural meaning then it does so in virtue of being an instance of a geometrical figure. But if it has non-natural meaning then it can be inherently general, a drawing of, say, an angle without being a drawing of any

---

[14] Jody Azzouni makes a related suggestion in his [2004]. He writes that "[T]he proof-relevant properties of diagrammatic figures . . . are not the actual (physical) properties of diagrammatic figures, but conventionally stipulated ones the *recognition* of which is *mechanically executable* . . . [T]he situation is exactly the same with games: Pieces are conventionally endowed with powers (e.g., to 'take' other pieces under certain circumstances) . . . [A]ll powers a game piece is stipulated to have are mechanically recognizable" (ibid.: 125).

angle in particular. The geometer draws some lines to form a rectilineal angle, in order, say, to show that an angle can be bisected. The geometer does not mean or intend to draw an angle that is right, or obtuse, or acute. He means or intends merely to draw an angle. Hence he says (I.9) that the angle he draws is "the given rectilineal angle," not also that it is right, or obtuse, or acute. That which he draws, regarded as something having natural meaning, will necessarily be right, or acute, or obtuse; but regarded as having non-natural meaning, it will be neither right nor acute nor obtuse. It will simply be an angle. One can no more infer on the basis of the drawing so conceived that the angle in question is (say) obtuse, based on how it looks, than in Grice's case one could infer (say) that Mr. Y has put on a little weight recently, based on how the drawing of him looks. A drawing of Mr. Y could have as its non-natural meaning that he has put on weight recently, just as in the context of a Euclidean demonstration a drawing of an angle can have as its non-natural meaning that it is (say) obtuse, but that would require, in both cases, that quite different intentions be in play.

If the figures drawn in a Euclidean diagram have non-natural rather than natural meaning, then they can, by intention, be essentially general. It seems furthermore clear that they would in that case function as icons in Peirce's sense, rather than as symbols or indices, because they would in some way *resemble* that which they signify.[15] But the resemblance is not, at least not merely and not directly, a resemblance in appearance. As Peirce notes, "[M]any diagrams resemble their objects not at all in looks; it is only in respect to the relations of their parts that the likeness consists" (Peirce [1931]: 282). We can, then, think of a drawn figure in Euclid as an icon that (though it may also resemble its object in appearance) signifies by way of a resemblance, or similarity, in the relations of parts, that is, in virtue of a homomorphism. For example, on such an account a drawn circle serves as an icon of a ge-

---

[15] Peircean icons can have either natural or non-natural meaning. In particular, individual instances of geometrical figures can be icons of the relevant sorts of things; they have in that case natural meaning. A drawn circle regarded as an instance of a circle is an icon of a circle that has natural meaning because it so functions independent of anyone's intention that it do so. But a drawn circle can also function as an icon with non-natural meaning. In that case it can be essentially general, an icon of a circle not further specified.

ometrical circle not in virtue of any similarity in appearance between the two but because there is a likeness in the relationship of the parts of the drawing, specifically in the relation of the points on the drawn circumference to the drawn center, on the one hand, and the relation of the corresponding parts of the geometrical figure, on the other.

A drawn circle is roughly circular; it looks like a circle just as a dog looks like a dog. But a dog looks like a dog because it *is* a dog, that is, a particular instance of doggy nature (as we can think of it). A drawn circle, I have suggested, can look like a circle for *either* of *two* reasons. It can look like a circle for the same reason that a dog looks like a dog, namely, because it is a circle, a particular instance of circle nature. Or it can look like a circle because it is a icon with non-natural meaning that is intended to resemble a circle first and foremost in the relation of its parts. Because what it is an icon of is circle nature, and because what is essential to a circle's being a circle is that all points on the circumference are equidistant from the center, and it is this relationship of parts that is to be iconically represented, the icon itself comes to look roughly circular. The appearance of circularity is induced in this case by the intended higher order resemblance rather than being something that is there in any case (as circularity is there in any case in a drawing of a particular instance of a circle).

Now if drawings in Euclid function as icons with non-natural meaning, this would explain why Euclid never mentions either straightedge or compass—which one might expect him to if it was an instance that was to be drawn. If one wishes to draw a particular straight line then the best way to do that is with a straightedge, and similarly in the case of a circle: If one wants to draw a particular circle then one is best off using a compass. If, however, one's aim is to draw something with non-natural meaning, in particular, an icon, say, of a line or circle, then all that matters is that one's drawing is able to produce the desired effect, to convey one's intention. In this case there is no reason at all to mention some particular means of producing the drawing precisely because the drawing, to serve the purpose it is to serve need not *look* very much like that for which it is an icon. This is not true of an instance: An instance ought as far as possible to look like what it is. It follows that instances are harder to draw than icons, and this is in fact generally true. It is, for instance, much easier to draw stick human figures and "smiley" faces, which are of course inherently general, than it is to draw someone in particular, the way some particular person actually looks. Even very small children can do the former; most of us even as adults cannot do the latter very well at all.[16] There is no need, then, for mechanical aids in drawing Euclidean diagrams if those diagrams function as has been suggested here, as essentially general icons with non-natural meaning.

More interestingly, the idea that diagrams are iconic and general by intention would also explain why ancient Greek mathematicians almost never make the sorts of mistakes they would be expected to make if their reasoning were based instead on an instance. If one were reasoning about an instance, it would be critical carefully to distinguish between what can and cannot be inferred on the basis of one's consideration of that instance. And this ought, in principle, to be quite difficult. It is, for instance, much harder to learn what a particular person looks like (so as to be able to re-identify that person) from a chance photograph than it is to find this out from, say, a caricature because in the caricature the work of discovering what are the salient and characteristic features of the person's appearance has already been done. Similarly, if a Euclidean drawing were an instance (with natural meaning), it would be hard to distinguish between what Manders has called co-exact features, ones on the basis of which an inference can be made, and exact features, which have no implications for one's course of reasoning (see Manders [1996]: 392–3). But in fact, all the evidence suggests that this is not hard at all. As Mueller remarks regarding the familiar diagram-based "proof" that all triangles are isosceles, "perhaps a 'pupil of Euclid' might stumble on such a proof; but probably he, and certainly interested mathematicians, would have no trouble in figuring out the fallacy on the basis of intuition and figures alone. And in the history of Euclidean geometry no such fallacious arguments are to be found. There are indeed many instances

---

[16] Notably, some very young autistic children lacking essentially all linguistic ability are able to draw very realistically, that is, things as they actually look, an ability that deteriorates as they learn language. This suggests that we generally draw not what things look like but instead something essentially general, something more like an icon of a concept. See Selfe [1977].

of tacit assumptions to be made, but these assumptions were always true. In Euclidean geometry . . . precautions to avoid falsehood are really unnecessary" (Mueller [1981]: 5). This is unsurprising if the figures in the diagram function not as instances but as icons (with non-natural meaning) that are inherently general. The features that Manders identifies as exact are no more there in the diagram than information about the relative size of human limbs is there in a stick figure.

We saw that drawings of geometrical figures conceived as icons (with non-natural meaning) can resemble particular instances in appearance in virtue of the fact that there is a resemblance in the relation of the parts. A drawn circle, intended as an icon of a circle, can look very much like a particular instance of a circle. It is easy, then, based on such an appearance, to think that the drawing is such an instance, not such an icon at all. There is good reason nonetheless for thinking that in Euclidean demonstrations the diagram functions iconically (i.e., non-naturally) rather than to provide an instance about which to reason. First, as Manders has noted, the latter idea "seems incompatible with the use of diagrams in proof by contradiction" (Manders [1996]: 391). To demonstrate, for instance, that a circle does not cut a circle at more than two points, one first sets out that two circles do cut at more than two points, say, at four. This is not a situation that can obtain; there is no instance to be drawn. And yet the diagram is drawn (as we will see). If the diagram is read instead as a Peircean icon with non-natural meaning there is no difficulty. What the diagram means non-naturally, the content it exhibits, is exactly what one aims to show is impossible, that a circle cuts a circle at more than two points.

Nor is this the only sort of case in which it is impossible to draw an instance. We know, because Euclid tells us in the opening section of the *Elements*, that a point is that which has no parts, and that a line segment is a length that has no breadth (the extremities of which are points). Such entities are clearly not perceptible; there is nothing that a thing with no parts, or a length with no breadth, looks like. It follows directly that there is no way to draw an instance of either a point or a line. On the other hand, such things, and their relations one to another, can be iconically represented. A drawn line length, for example, can represent (be intended iconically to signify) a line with endpoints. A drawn dot can represent (be intended iconically to signify) a point. A drawn circle is, again, a slightly different case because drawn circles do look roughly circular; that is, there is a look that geometrical circles can be said to have. But as in the case of Grice's drawing, the role of a drawn circle in the context of a Euclidean demonstration is not, on the account we are pursuing, that of an instance but instead that of an icon with non-natural meaning, one that is intended to represent the relation of the points on the circumference to the center, the fact that those points are equidistant from the center (whether or not in the figure as drawn the points on the circumference *look* equidistant from the center). All that is represented in a Euclidean straight line (conceived as an icon) is a breadthless length lying evenly with the points on itself, that is, a certain relationship between the line and the points that may be found on it; and similarly, all that is represented in a Euclidean circle (similarly conceived) is the relation of the points on its circumference to the center. In each case, what is important for the cogency of the demonstration is not what the figure looks like but instead what is intended by it whether as set out in Euclid's definitions and postulates, or as stipulated by the particular problem or theorem in question. The similarity in appearance (say, between the icon of a circle and an actual instance of something circular) does help to convey the intended meaning, just as it does in the example of Grice's drawing; nevertheless, what is meant is carried by the intention, not by the similarity in appearance. It is for just this reason that the demonstration can be seen to be essentially general throughout.[17] What one draws in Euclidean diagrams are not pictures of

---

[17] This is, furthermore, consistent with the final step in the demonstration. As Ian Mueller explains, "[I]n ancient logic the *sumperasma* is the conclusion of an argument. In the Elements, however, the *sumperasma* is not so much a result of an inference as a summing up of what has been established" (Mueller [1974]: 42). As he notes in his [1981], "[T]he word sumperasma [which marks the last stage of a Euclidean demonstration] can . . . mean 'completion' or 'finish' . . . [T]he *sumperasma* merely sums up what has taken place in the proof . . . It merely completes the proof by summarizing what has been established" (ibid.: 13). So conceived, the move from a claim involving letters to the explicitly general claim at the end of a Euclidean demonstration is essentially a move from an implicit generality, such as that a horse is warm-blooded (inferred perhaps from the claim that a horse is mammal together

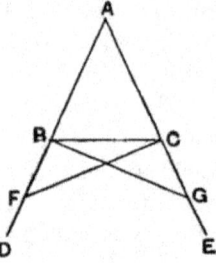

Figure 2

geometrical objects but the relations that are constitutive of the various kinds of geometrical entities involved. As the point might be put, a Euclidean diagram does not *instantiate* content but instead *formulates* it.

3 Diagrammatic Reasoning in Euclid's Elements

We have seen that Euclid's system is best thought of on analogy with a system of natural deduction as it contrasts with an axiomatization. And I have argued that figures in Euclidean diagrams do not provide instances but instead should be read as having non-natural meaning and to function, in particular, as icons. Diagrams so conceived are constitutively general. What remains to be shown is that the reasoning involved in a Euclidean demonstration is not merely diagram-based but instead diagrammatic.

It is clearly true that in a Euclidean demonstration at least some steps in the chain of reasoning are in some way licensed by the diagram. Consider, for example, proposition I.5, that in isosceles triangles the angles at the base are equal to one another, and if the equal straight lines be produced further, the angles under the base will be equal to one another. The diagram is shown in Figure 2.[18] It is then argued that because AF is equal to AG and AB to AC, the two sides FA,

---

with the fact that mammals are warm-blooded) to the explicit generality that all horses are warm-blooded.

[18] The image is taken from Euclid: vol. 1, 251.

AC are equal to the two sides GA, AB, and they contain a common angle FAG. This inference, from the two equalities, AF to AG and AB to AC, to the double conclusion, first that the two sides FA, AC are equal to the two sides GA, AB, and also that both those same pairs of sides contain the angle FAG, might be thought to be diagram-based in the following sense. Having, first, drawn an isosceles triangle with sides AB and AC equal, and then having produced the straight lines BD and CE by extending (respectively) AB and AC, and taken F on BD at random and cut AG off from AE to equal AF, one has thereby inevitably produced not only FA, AC equal to GA, AB, which is already more or less given, but also the angle FAG, which, as inspection of the diagram shows, is identical both to the angle FAC and to the angle GAB. It is no more possible to construct the required diagram without realizing this identity of angles than it is to embed one circle in a second circle and the second in a third without thereby embedding the first in the third. (This is of course the principle behind Euler diagrams, which can be used to exhibit the validity of various valid syllogistic forms of reasoning, here *barbara*.) So, just as one can infer on the basis of an Euler diagram that some conclusion follows, so one can infer on the basis of the above diagram that the relevant claims are true. The diagram-based inference is, in both cases, valid, that is, truth-preserving, and it is truth-preserving precisely because there exists a homomorphism, a higher level formal identity, that relates relations among elements in the diagram with relations among the entities represented by those elements.

Reasoning using Euler diagrams provides a paradigm of what we might mean by diagram-based inference. In that case, one draws icons of things of a given sort, in particular, circles to represent collections of objects, and furthermore draws them in spatial relations that are homomorphic to relations of inclusion and exclusion over collections. If, for example, all dogs are mammals then one draws the circle representing the collection of dogs inside the circle representing the collection of mammals. If no fish are mammals then one draws the circle representing the collection of fish wholly separate from that representing the collection of mammals, without any overlap at all. Then one can just read off the diagram what the relationship is between the collection of dogs and the collection of fish: The relevant icons are wholly disjoint so one infers that no dogs are fish. The inference, the passage from the premises about dogs, mammals, and fish, to the conclusion about dogs and fish is licensed by the diagram one draws. It is diagram-based. Do diagrams Euclidean demonstrations function in the same way as the basis for an inference? We will see that they do not.

The first indication that diagrams in Euclid function differently from Euler diagrams is the fact that although Euler diagrams can make something that is implicit in one's premises explicit, they cannot in any way extend one's knowledge. Euclidean demonstrations, by contrast, do seem clearly to be fruitful, real extensions of our knowledge. And the reason they are is surely connected to the fact that objects can pop up in a Euclidean diagram. We find out that an equilateral triangle can be constructed on a straight line only because a triangle pops up in the course of the construction. Before that point there is nothing whatever about any triangles in anything we have to work with in the demonstration. Quite simply, there is no sense in which an equilateral triangle is *implicit* in a line, even given Euclid's axioms, postulates, and definitions. Nevertheless, as I.1 shows, such a triangle *is* there *potentially*: given what Euclid provides us, we can construct an equilateral triangle on a given straight line. The triangle just pops up when we draw certain lines. And this is true generally in ancient Greek mathematics: One achieves something new largely in virtue of this pop-up feature of its diagrams. To understand the role of diagrams in Euclidean demonstrations, then, we need to understand these pop-up objects.

In a Euclidean demonstration, what is at first taken to be, say, a radius of a circle is later in the demonstration seen as a side of a triangle. But how *could* an icon of one thing become an icon of another? How, for example, could an icon of a radius of a circle *turn into* an icon of a side of a triangle? Certainly nothing like this happens in the course of reasoning about an Euler diagram. The iconic significance of the drawn circles is fixed and unchanging throughout the course of reasoning. If the lines one draws in a Euclidean diagram had this same sort of iconic significance, the reasoning would clearly stall because in that case no new points or figures *could* pop up. It is also obvious that nothing could possibly be *both* an icon of (say) a radius of a circle and an icon of a side of a triangle at once, because these are incompatible. An icon of a radius essentially involves reference to a circle; radii are and must be radii of circles. An icon of a side of a triangle makes no reference to a circle. So nothing could at once be an icon of both. But one and the same thing *could serve* now (at time t) as an icon of a radius, and now (at time $t' \neq t$) as an icon of a side. The familiar duck–rabbit drawing is just such a drawing; it is a drawing that is an icon of a duck (though of course no duck in particular) when viewed in one way and an icon of a rabbit (no one in particular) when viewed in another.

Some drawings, such as the duck–rabbit, can be seen in more than one way. It would seem, then, that we can say that in the diagram of I.1 one sees certain lines now as icons of radii, as required to determine that they are equal in length, and now as icons of sides of a triangle, as required in order to draw the conclusion that one has drawn an equilateral triangle on a given straight line.[19] The cogency of the reasoning clearly requires both perspectives. But if it does then the Euclidean diagram is functioning in a way that is very different from the way an Euler diagram functions. Much as in the case of the duck–rabbit drawing, and by contrast with an Euler diagram, various collections of lines and points in a Euclidean diagram are icons of, say, circles, or other particular sorts of geometrical figures, *only when viewed a certain way*, only when, as Kant would think of it, the manifold display (or a portion of it) is synthesized under some particular concept, say, that of a circle, or of a triangle.[20] The point is not that the drawn lines underdetermine what is iconically represented, as if they have to be supplemented in some way. It is that the drawing as given, as certain marks on the page, has (intrinsically)

---

[19] Euclid's definitions would seem to play a crucial role here: Only if one can see in the diagram that the definition is satisfied can one find the relevant object in it.

[20] This idea, that a system of signs may function not by way of combinations of primitive signs that have designation independent of any context of use, but instead by appeal to primitive signs that designate only given a context of use, is appealed to also in Sun-Joo Shin's ([2002]) reading of Peirce's notation for his alpha logic. In my [2005], I argue that Frege's notation functions in an essentially similar way. The primitive signs of the language express senses independent of their occurrence in a sentence but designate only in the context of a sentence and relative to an analysis, relative, that is, to a way of regarding the whole two-dimensional array of signs.

the *potential* to be regarded in radically different ways (each of which is fully determinate albeit general, again, as in the duck–rabbit case). It is just this potential that is actualized in the course of reasoning, as one sees lines now as radii and now as sides of a triangle, and which begins to explain the enormous power (and beauty) of Euclidean demonstrations as against the relative sterility (and lumpishness) of Euler diagrams.

In Euclidean diagrams, geometrical entities, points, and figures, pop up as lines are added to the diagram. Cut a line AB with another line CD and up pops a point E as the point of intersection, as well as four new lines AE, BE, CE, and DE; take the diagonal of a square and up pop two right triangles, and so on. We do not find this surprising in practice; indeed, one need have no self-conscious awareness that it is happening as one follows the course of a Euclidean demonstration. Nevertheless, as we have seen, it clearly *is* happening, and the cogency of the demonstration essentially depends on it. Such pop up objects, I have suggested, depend in turn on our capacity to "re-gestalt" various collections of lines, to see them now one way and now another. And what this shows is that it is not the lines themselves that function as icons (even in light of one's intention that they be so regarded) but only the lines *when seen from a particular perspective*, when viewed in one way rather than another equally possible way.

Consider again our two crossed lines AB and CD that cut at E, and suppose that they are functioning as icons independent of any perspective taken on them. One could then argue that the point E can belong to only one of the four segments AE, BE, CE, or DE, leaving the other three without an endpoint at one end. This would be a perfectly reasonable inference if we were dealing with a simple icon, one whose significance was fixed independent of any perspective taken on it, because it is true that if you divide a (dense) line by a point then that point can be the endpoint of only one of the two line segments, leaving the other to approach it indefinitely closely (because the line is dense), without having it as its endpoint. Suppose now that we offered this argument to an ancient Greek geometer. How would he respond? He would laugh us away, much as he laughs away those who would "[attempt] to cut up the 'one'" by pointing out that the line drawn to iconically represent the unit can surely be divided.[21] The geometer laughs because one is in that case taking the diagram the wrong way, because one fails to understand how it works as a diagram.

When two lines AB and CD cut at E, only one point pops up and it is the endpoint of all four lines AE, BE, CE, and DE, which also just pop up with the cut. And they can do because E is the endpoint of any one of these lines only relative to a way of regarding it, much as certain lines in the duck–rabbit drawing are ears, or a duckbill, only relative to a way of regarding those lines. Another example makes the same point. Suppose that I claimed on the basis of the drawing of a straight line segment ABCD that, contrary to what Euclid holds, two straight lines *can* have a common segment: AC and BD have BC in common. Again the geometer would laugh—and rightly so. Of course one can see the drawn line as iconically representing AC, or BD, or AD, or BC, and indeed one may need so to regard it in the course of a demonstration; but that does not show that the lines themselves have a common segment. This would be required only if the drawn line functioned iconically to represent these possibilities independent of any perspective that was taken on the drawing. Between two points there is only one straight line, not many; two points can seem to have a common segment (as above) only if one, perversely, mistakes the way the diagram functions.

I have been arguing that the lines in a Euclidean diagram, like the lines in a duck–rabbit drawing, function as icons of various sorts of geometrical objects only relative to a perspective that is taken on them. But those diagrams have a further feature as well, one that is not found either in the duck–rabbit drawing or in an Euler diagram, a feature that is essential to the fruitfulness of Euclidean demonstrations. Euclidean diagrams are distinguished in having three levels of articulation as follows. At the lowest level are the primitive parts, namely, points, lines, angles, and areas. At the second level are the geometrical objects we are interested in, those that form the subject matter of geometry, all of which are wholes of those primitive parts. At this level we find points as endpoints of lines, as points of intersection of lines, and as centers of circles; we find angles of various sorts that are limited by lines that are

---

[21] Plato, *Republic Book VII*: 525e.

also parts of those angles; and we find figures of various sorts. A drawn figure such as (say) a square has as parts: four straight line lengths, four points connecting them, four angles all of which are right, and the area that is bounded by those four lines. Of course, in the figure as actually drawn, the lines will not be truly straight or equal, and they will not meet at a point; the angles will not be right or all equal to one another. But this does not matter because the drawn square is not a picture or instance of a square but instead an icon of a square, one that formulates certain necessary properties of squares. At the third level, finally, is the whole diagram, which is not itself a geometrical figure but within which can be discerned various second-level objects depending on how one configures various collections of drawn lines within the diagram. What we will see is that these three levels of articulation, combined with the possibility of a variety of different analyses or carvings of the various parts of the diagram, account for the ways in which one can radically reconfigure parts of intermediate wholes into new (intermediate) wholes in the course of a Euclidean demonstration, and thereby demonstrate significant and often surprising geometrical truths.

As already noted, a Euclidean demonstration comprises a diagram and some text, most importantly for our current purposes, the *kataskeue* (construction) and the *apodeixis*. The *kataskeue* provides information about the construction of the diagram and is governed by what can be formulated in the diagram as legitimated by the postulates and any previously demonstrated problems. The *apodeixis*, which is governed by what can be read off the diagram as legitimated by the definitions, common notions, and previously demonstrated theorems, should be read, on our account, *not* as the medium of the proof but instead as providing instructions regarding how various portions of the constructed diagram are to be read, construed, or analyzed, that is, how they are to be carved up in the course of the demonstration. It is the *diagram* that is the site of reasoning, on our account, not the accompanying text.

Consider one last time the first demonstration in Euclid, that on a given straight line an equilateral triangle can be constructed. The diagram is displayed, again, in Figure 3. According to the reasoning already rehearsed, we know that AC = AB because A is the center of

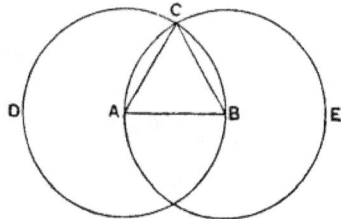

Figure 3

circle CDB, and that BC = BA because B is the center of circle CAE, hence that AC = BC = AB, on the basis of which it is to be inferred that ABC is an equilateral triangle constructed on the given line AB, which was what we were required to show. Our task concerning this little chain of reasoning is to understand how, based on a claim about radii of circles, one might infer, even given the diagram, something about a triangle. Were the reasoning merely diagram-based, that is, were it reasoning *in* (natural) language but, at least in some cases, *justified by* what is depicted in the diagram then the problem seems intractable. No mere diagram of some drawn circles, however iconic, can justify an inference from a claim about the radii of circles to a claim about a triangle. If, on the other hand, we take the diagram not merely as a collection of icons of various geometrical entities but instead as a display with the three levels of articulation outlined above, and so as a collection of lines various parts of which can be configured and reconfigured in a series of steps (as scripted by the *apodeixis*), the solution of our difficulty is obvious. One considers now one part of the diagram in some particular way and now another in some (perhaps incompatible, but perfectly legitimate) way as one makes one's passage to the conclusion. In particular, a line that is at first taken iconically to represent a radius of a circle is later taken iconically to represent a side of a triangle. It is only because the drawing can be regarded in these different ways that one can determine that certain lines are equal, because they can be regarded as radii of a single circle, and then *conclude* that a certain triangle is equilateral, because its sides are equal in length. The demonstration is fruitful, a

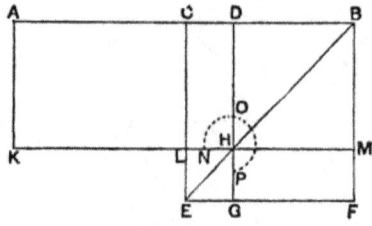

Figure 4

real extension of our knowledge, for just this reason: Because we were able to take a part of one whole and combine it with a part of another whole to form an utterly new and hitherto unavailable whole, we were able to discover something that was simply not there, even implicitly, in the materials with which we began.

4 More Examples

Consider now the Euclidean demonstration of Proposition II.5: If a straight line be cut into equal and unequal segments, the rectangle contained by the unequal segments of the whole together with the square on the straight line between the points of section is equal to the square on the half. The diagram is shown in Figure 4.[22] We letter it for ease of reference in our written discourse; were the presentation an oral one would need only to point to relevant aspects of the drawn diagram. To understand this diagram, that is, to know how to "read" it, configure its parts (at least to begin with), we need to know the provenance of its various parts. We are given that line AB is straight, and that it is cut into equal segments by C and into unequal segments by D. We know by the construction that CEFB is a square on CB, which we know from proposition I.46 is constructible, and that DG is a straight line parallel to CE and BF, KM parallel to AB and EF, and AK parallel to CL and BM, shown to be constructible in proposition I.31. That is, already we see here the various ways the drawn lines are taken to figure in various iconic representations of geometrical figures. CB, for example, is first

---

[22]The image is taken from Heath [1956]: vol. 1, 382.

taken to be the half of line AB, but then as a side of a square. BM is a line length equal in length to DB but also a part of the line BF, which is another side of that same square. It is in virtue of these various relations of parts iconically represented in the drawn diagram, and of these reconfigurings of parts, that the diagram can show that the theorem is true. It does so in a series of six steps scripted by the *apodeixis*:

(1) 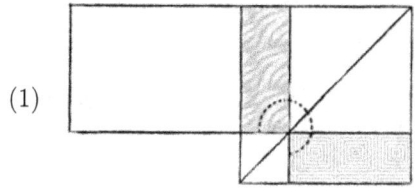 These (shaded) areas are equal, as is shown by Prop. I.43: The complements of a parallelogram about the diameter are equal to one another.

(2) 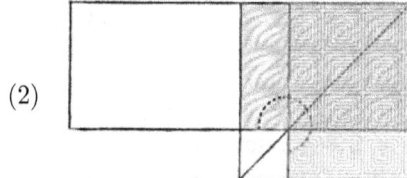 It follows from (1) that these areas are equal, because the same has been added to the same.

(3) 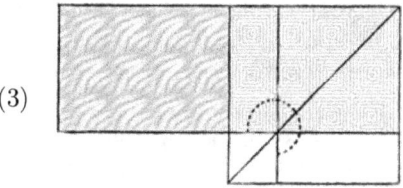 These areas are known to be equal on the basis of what we know about the relationships that obtain among the lines that form the boundaries of the two shaded areas.

(4) 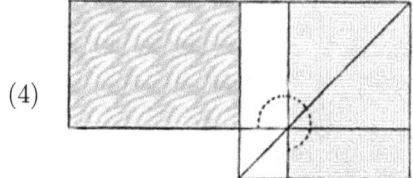 It follows from (2) and (3) that these areas are equal because things equal to the same are equal to each other.

(5) 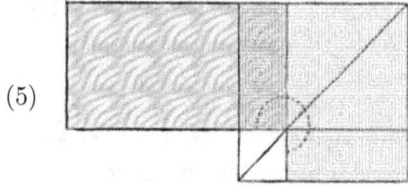 It follows from (4) that these areas are the same because the same has been added to equals.

(6) 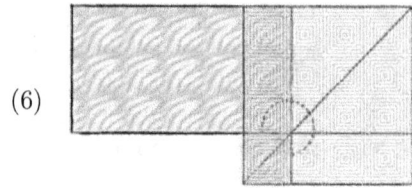 It follows from (5) that these areas are equal because the same has been added to equals.

But that is just what we wanted to show: that the square on the half is equal to the rectangle contained by the unequal segments plus the square on the line between the points of section. By construing various aspects of the diagram in these various ways in the appropriate sequence one comes to see that the theorem is, indeed, must be, true. But if that is right, then (again) it is the diagram that is the site of reasoning in Euclid, not the text. The text, on this account, is merely a script to guide one's words, and thereby one's thoughts, as one walks oneself through the demonstration.[23] The demonstration is not in the words of the *apodeixis*; it is in what those words help us to see in the diagram.[24]

---

[23] Think of doing a calculation in Arabic numeration, say, working out the product of thirty-seven and forty-two. One first writes the two Arabic numerals in a particular array, one directly above the other, then, in a series of familiar steps performs the usual calculation. As one writes down the appropriate numerals in the appropriate places, one may well talk to oneself ('Let's see; two times seven is fourteen, so . . . '). Nevertheless, the calculation is not in the words one utters, if one does utter any words, even silently (and one need not), but in the written notation itself. Just imagine trying to do the calculation without appeal to Arabic numeration! One calculates *in* Arabic numeration. In much the same sense, I am suggesting, one reasons *in* a Euclidean diagram.

[24] That the words of the demonstration in Euclid are merely a record of the speech of someone presenting the demonstration to an audience is also indicated by the way the written text is presented in Euclid: "unspaced, unpunctuated, unparagraphed, aided by no symbolism related to layout." "Script," in Euclid, "must be transformed into pre-written language, and then be interpreted through the natural capacity for seeing [better: hearing] form in [spoken] language" (Netz [1999]: 163).

(This would perhaps be even more obvious if we considered someone trying to discover a demonstration of some proposition. The inscribing would not be of words but of diagrams.)

As this simple demonstration again shows, in order to follow, that is, to understand, a Euclidean demonstration, one must be able to see various drawn lines and points now as parts of one iconic figure and now as parts of another. The drawn lines are assigned very different significances at different stages in one's reasoning; they are *interpreted* differently depending on the context of lines they are taken, at a given stage in one's reasoning, to figure in. It is just this that explains the fruitfulness of a Euclidean demonstration, the fact that its conclusion is, as Kant would say, synthetic *a priori*. In Euclid, the desired conclusion is contained in the diagram, drawn according to one's starting point and the postulates, not merely implicitly, needing only to be made explicit (as in a deductive proof on the standard construal or in an Euler diagram), but instead only potentially. The potential of the diagram to demonstrate the conclusion is made actual only through an actual course of reasoning in the diagram, that is, by a series of successive refigurings of what it is that is being iconically represented by various parts of the diagram. Parts of wholes must be taken apart and combined with parts of other wholes to make quite new wholes. And this is possible, first, in virtue of the three levels of articulation in the diagram, and also because the various parts of the diagram signify geometrical objects only relative to ways of regarding those parts. A given line must actually be construed now as an iconic representation of (say) a part of another line and now as an iconic representation of a side of a square, if the demonstration is to succeed. The diagram, more exactly, its proper parts, must be actualized, now as this iconic representation and now as that, through one's construal of them as such representations, if the result is to emerge from what is given. Only a course of thinking through the diagram can actualize the truth that it potentially contains.[25]

Another example, this time of a reductio proof, is the demonstration of III.10, that a circle does not cut a circle at more points than two. The diagram (again lettered to enable us to explain in the written text what it involves) is shown in Figure 5.[26] The following is encoded, or

---

[25] We might, then, translate '*apodeixis*' not as 'proof' but as 'reasoning.'
[26] The image is taken from Heath [1956]: vol. 2, 23.

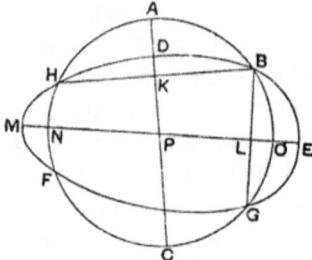

Figure 5

formulated, iconically in the diagram. First, we have by hypothesis (for reductio) that the circle ABC cuts the circle DEF at the points B, G, F, and H; and here (again) it is especially obvious that we do not *picture* the hypothesized situation, which is of course impossible, but instead formulate in the diagram the content of that hypothesis. BH and BG are drawn (licensed by the first postulate) and are then bisected at K and L respectively. (We know that we can do this from I.10.) KC is then drawn at right angles to BH, and LM at right angles to BG. (We know from I.11 that we can do this.) Both KC and LM are extended, KC to A and LM to E, as permitted by the second postulate. What we have then are, first, two straight lines BH and BG, both of which are chords to both circles; that is, we can see BH as drawn through ABC or as drawn through DEF, and similarly for BG. And we also know that AC bisects HB and is perpendicular to it, and similarly for ME and BG. Now the reasoning begins:

(1) 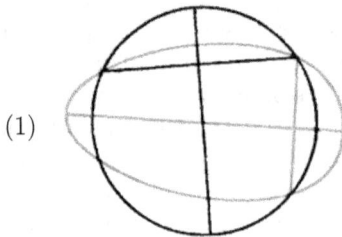 The center of the highlighted circle is on the highlighted bisecting line, by III.1 porism.

(2) 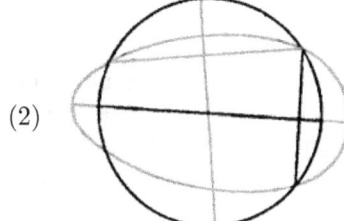 The center of the highlighted circle is on the highlighted bisecting line, again by III.1 porism.

(3) 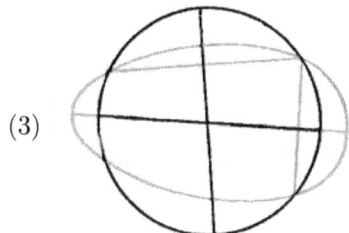 As the diagram shows, the two bisecting lines meet at only one point; so this point must be the center of the circle (because it is the only point that is on both lines).

(4) 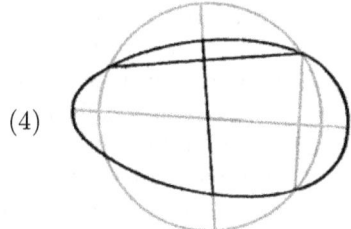 The center of the highlighted circle is on the highlighted bisecting line, by III.1 porism.

(5) 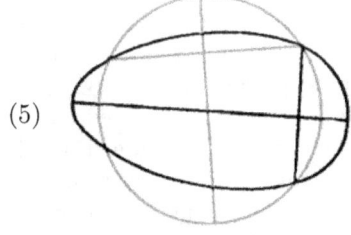 Again, the center of the highlighted circle is on the highlighted bisecting line, by III.1 porism.

(6) 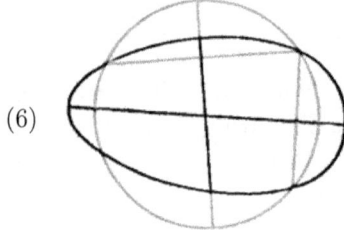 The two bisecting lines meet at only one point and so that point must be the center of the circle (because, again, that is the only point on both lines known to contain the center).

(7) 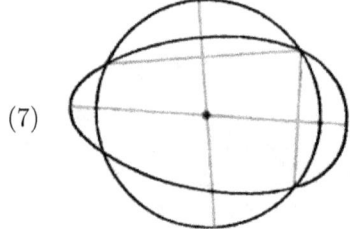 The two circles have the same center because one and the same point that was shown first (3) to be the center of the circle ABC was shown also (6) to be the center of DEF.

But we know from III.5 that if two circles cut one another then they will not have the same center. Our hypothesis that a circle cuts a circle at four points is false. Furthermore, because the demonstration involved only three of the four points presumed to cut the two circles, it clearly is cogent no matter how many points one supposes the two circles to cut at. Any number more than two will lead to contradiction. It follows that a circle does not cut a circle at more points than two.

In this demonstration we are concerned with lines, circles, more exactly, the circumferences of circles, and their interrelationships. All that we assume about the various figures is what we are told about them in the construction, and all that we infer at each stage is what we know follows from what we know about the figures that are taken at that stage to be iconically represented in the relevant part of the diagram. By focusing on the various parts of the diagram, now this part and now that, and conceptualizing these parts appropriately, then applying what one knows to the relevant figure thereby iconically represented, one can easily come to realize that, given what is known and assumed, the two circles have one and the same center. But this is absurd given that they cut one another. The assumption that a circle can cut a circle at more than two points must be rejected. It is in just this way that one reasons by reductio in a Euclidean diagram, by assuming in a diagram what one aims to show is false and then showing by a chain of strict and rigorous reasoning in the diagram that that assumption leads to a contradiction.

5 Conclusion

We have seen that diagrams in Euclid are not merely images or instances of geometrical figures but are instead icons with Gricean non-natural meaning. As such, they are inherently general. But as we also saw, this alone is not sufficient to explain the workings of a Euclidean demonstration. In particular, Euclidean reasoning cannot be merely diagram-based, even where the diagram is conceived iconically, because then no account can be given of the pop-up objects that are essential to the cogency of Euclidean demonstrations. Euclidean reasoning is instead properly diagrammatic; one reasons *in* the diagram in Euclidean geometry, actualizing at each stage some potential of the diagram. This is furthermore made possible, we have seen, by the nature and structure of the signs that make up that system. Because Euclidean diagrams are collections of primitive signs with three levels of articulation, the parts can be variously construed in systematic ways to mean, iconically represent, now this geometrical figure and now that. The conclusion that one draws on the basis of a Euclidean demonstration is, for just this reason, contained in one's starting points only potentially. The steps of the demonstration *must* be taken to actualize that conclusion. A Euclidean demonstration is not, then, diagram-based, its inferential steps licensed by various features of the diagram. It is properly diagrammatic. One reasons *in* the diagram, in Euclid, that is, through lines, *dia grammon*, just as the ancient Greeks claimed (Netz [1999]: 36).

The picture of Euclid's *Elements* that has emerged is, then, very different from that with which we began. First, a Euclidean demonstration, as we have come to understand it, is not a *proof* in the standard sense at all; it is not, that is, a sequence of sentences some of which are premises and the rest of which follow in a sequence of steps that are deductively valid, or diagram-based. Indeed, the demonstration does not lie in *sentences* at all. It lies in a certain activity, in a course of reasoning focused directly on a diagram. And this activity is needed, we have seen, because it alone can actualize the potential of the diagram to yield something new. The diagram, the actual marks on the page (or some appropriate analogue), must be seen now this way and now that if the result that is wanted is to be achieved. But if so, then Euclid's geometry is not an axiomatic system, that is, a collection of primitive and derived sentences; it does not involve reasoning about an instance (of a geometrical figure); and it is not merely diagram-based. It is instead a mode of mathematical inquiry, a mathematical practice that uses diagrams to explore the myriad discoverable necessary relationships that obtain among geometrical concepts, from the most obvious to the very subtle.

Acknowledgments

This paper is a much longer version of the paper I presented at the PMP 2007 conference on the philosophy of mathematical practice in Brussels. I have been helped in preparing this version by comments Ken Manders made at the time.

Bibliography

Azzouni, J. [2004]. "Proof and Ontology in Euclidean Mathematics." *New Trends in the History and Philosophy of Mathematics.* Ed. by T.H. Kjeldsen; S.A. Pederson; and L.M. Sonne-Hansen. Odense, Denmark: University Press of Southern Denmark.

Dipert, R.R. [1996]. "Reflections on Iconicity, Representation, and Resemblance: Pierce's Theory of Signs, Goodman on Resemblance, and Modern Philosophies of Language and Mind." *Synthese* **106**. 373–97.

Grice, H.P. [1957]. "Meaning." *Philosophical Review* **66**. 377–88.

Heath, T.L. [1956]. *The Thirteen Books of the Elements.* 2nd ed. First published in 1908. Toronto: Dover.

Lachterman, D.R. [1989]. *The Ethics of Geometry: A Genealogy of Modernity.* New York and London: Routledge.

Leibniz, G.W. [1969]. *Philosophical Papers and Letters.* Trans. by L.E. Loemker. Dordrecht: Reidel.

Macbeth, D. [2004]. "Viète, Descartes, and the Emergence of Modern Mathematics." *Graduate Faculty Philosophy Journal* **25**. 87–117.

— [2005]. *Frege's Logic.* Cambridge, MA: Harvard University Press.
Manders, K. [1996]. "Diagram Content and Representational Granularity." *Logic, Language, and Computation.* Ed. by J. Seligman and D. Westerståhl. Vol. I. Stanford: CSLI Publications and Stanford University Press.
— [2008]. "The Euclidean Diagram." *The Philosophy of Mathematical Practice.* Ed. by M. Paolo. Oxford: Oxford University Press.
Mueller, I. [1974]. "Greek mathematics and Greek logic." *Ancient Logic and its Modern Interpretation.* Ed. by J. Corcoran. Dordrecht and Boston: D. Reidel.
— [1981]. *Philosophy of Mathematics and Deductive Structure in Euclid's Elements.* Cambridge, MA and London: MIT Press.
Netz, R. [1999]. *The Shaping of Deduction in Greek Mathematics.* Cambridge: Cambridge University Press.
Peirce, C.S. [1931]. *Collected Papers of Charles Sanders Peirce.* Ed. by C. Hartshorne and P. Weiss. Vol. 2. Cambridge, MA: Harvard University Press.
Russell, B. [1956]. "Mathematical Logic Based on the Theory of Types." *Logic and Knowledge: Essays 1901–50.* Ed. by R.C. Marsh. Book first published in 1956. Article first published in 1908. London and New York: Routledge.
Ryle, G. [1950]. "'If,' 'So,' and 'Because'." *Philosophical Analysis: A Collection of Essays.* Ed. by M. Black. Ithaca, NY: Cornell University Press.
Selfe, L. [1977]. *Nadia: A Case of Extraordinary Drawing Ability in Children.* London: Academic Press.
Shin, S.-J. [2002]. *The Iconic Logic of Peirce's Graphs.* Cambridge, MA and London: Bradford Books and MIT Press.
Stein, H. [1988]. "*Logos*, Logic, and *Logistiké*: Some Philosophical Remarks on the Nineteenth Century Transformation of Mathematics." *History and Philosophy of Modern Mathematics.* Ed. by W. Aspray and P. Kitcher. Vol. XI. Minnesota Studies in the Philosophy of Science. Minneapolis, MN: University of Minnesota Press.

ESSAY 11

Observations on Sick Mathematics
Andrew Aberdein

> And you claim to have
> discovered this . . . from
> observations on *sick people*?
>
> (Freud [1990]: 13)

This paper argues that new light may be shed on mathematical reasoning in its non-pathological forms by careful observation of its pathologies. The first section explores the application to mathematics of recent work on fallacy theory, specifically the concept of an 'argumentation scheme': a characteristic pattern under which many similar inferential steps may be subsumed. Fallacies may then be understood as argumentation schemes used inappropriately. The next section demonstrates how some specific mathematical fallacies may be characterized in terms of argumentation schemes. The third section considers the phenomenon of correct answers which result from incorrect methods. This turns out to pose some deep questions concerning the nature of mathematical knowledge. In particular, it is argued that a satisfactory epistemology for mathematical practice must address the role of luck. Otherwise, we risk being unable to distinguish successful informal mathematical reasoning from chance. We shall see that argumentation schemes may play a role in resolving this problem too.[1]

Mathematical errors occur in many different forms. In his classic short treatment of the topic, E.A. Maxwell distinguished the simple MISTAKE, which may be caused by 'a momentary aberration, a slip in writing, or the misreading of earlier work,' from the HOWLER, 'an error which leads *innocently* to a *correct* result,' and the FALLACY, which 'leads by *guile* to a *wrong* but plausible conclusion' (Maxwell [1959]: 9). We might gloss this preliminary typology of mathematical error as correlating the (in)correctness of the result to the (un)soundness of the method, as in Table 1.

	True Result	**False Result**
Sound Method	Correct	Fallacy
Unsound Method	Howler	Mistake

Table 1: A preliminary typology of mathematical error

However, there are a number of problems with this picture. Most significantly, a truly sound method would never lead to a false conclusion. This paper addresses these problems, and seeks to derive some general lessons from the treatment of the fallacy, the howler, and related sources of confusion.

1 Fallacies

It might be supposed that mathematical fallacies could be defined very simply. If all mathematical reasoning is formal and deductive, then surely mathematical fallacies are merely invalid arguments? This definition has several shortcomings. Firstly, there are many invalid math-

---

[1] I am grateful to an anonymous referee for some of the phrasing in this paragraph.

ematical arguments which would not normally be described as mathematical fallacies. Secondly, much reasoning in mathematics is conducted informally. So a satisfactory account of mathematical fallacies must explain what is distinctive about formal fallacies, beyond their invalidity, and also address informal fallacies.

Most logic textbooks contain a chapter on informal fallacies. But this is typically no more than a catalogue of (not very) misleading arguments, which endures through successive editions, much as Oliver Wendell Holmes said of legal precedent, like 'the clavicle in the cat [which] only tells of the existence of some earlier creature to which a collar-bone was useful' (Holmes [1991]: 35). No attempt is made to provide an underlying theory which might explain their significance, their application, or their continued presence. However, in recent years fallacy theory has been reinvigorated. We shall see whether this new lease of life may be passed on to the study of the mathematical fallacy.

1.1 THE STANDARD TREATMENT

Most textbooks echo the 'standard treatment' of fallacy, as an argument which *'seems to be valid* but *is not* so' (Hamblin [1970]: 12). This definition may be traced back to Aristotle, for whom 'that some reasonings are genuine, while others seem to be so but are not, is evident. This happens with arguments as also elsewhere, through a certain likeness between the genuine and the sham' (Aristotle [1995]: 164a). The problem with the standard treatment, which has brought it into disrepute, is that we have no systematic account of the apparently subjective and psychologistic concept of 'seeming valid.'

This problem is perhaps especially acute for mathematics. As the philosopher and novelist Rebecca Goldstein has her fictitious mathematician Noam Himmel declare,

> [I]n math things are exactly the way they seem. There's no room, no *logical* room, for deception. I don't have to consider the possibility that maybe seven isn't really a prime, that my mind conditions seven to appear a prime. One doesn't—can't—make the distinction between mathematical appearance and reality, as one can—must—make the

> distinction between physical appearance and reality. The mathematician can penetrate the essence of his objects in a way the physicist never could, no matter how powerful his theory. We're the ones with our fists deep in the guts of reality. (Goldstein [1983]: 95)

On this account, Aristotle's 'likeness between the genuine and the sham' could never arise, since there would be no sham. This would rule out the possibility of mathematical fallacies, at least on the standard treatment.

However, the difference between appearance and reality which Himmel believes impossible is not that upon which the standard treatment rests. Himmel is concerned with radical scepticism about mathematical truth. He thinks it just could not happen that, despite our best efforts, we were fundamentally wrong. His position is persuasive, although not indisputable,[2] but it still leaves room for a weaker, error-based distinction between appearance and reality. Students and researchers alike frequently have the experience of some mathematical object appearing to be other than it turns out really to be—because they do not yet understand the matter. Himmel's argument might be paraphrased as the thesis that mathematical understanding guarantees mathematical truth. Once one understands the concept of prime number, one realizes that seven must be prime. This is not true in natural science—understanding the concepts of ether, caloric, or phlogiston does not make them real. But in both cases, if understanding is absent, truth may well be also. This leaves plenty of room for mathematical fallacies.

1.2 OTHER DISTINCTIONS

Before considering alternatives to the standard treatment we will examine some further distinctions which may be drawn amongst fallacies. Many theorists distinguish the SOPHISM, which is intended to deceive, from the PARALOGISM, an innocent mistake in reasoning. This distinction is sometimes made part of the definition of specific fallacies. But that seems to miss the point. Since a fallacy is a sort of argument, if the same argument is used, then the same fallacy should be committed.

---

[2]Descartes ([1996]: 14), for one, would dispute it. I am grateful to James Woodbridge for reminding me of this.

The speaker's intent should not make a difference. In fact, every fallacy can occur as either a sophism or a paralogism, since any fallacy can be used deliberately to deceive another, who, if he does not realize that he has been deceived, may guilelessly, if negligently, repeat the fallacy to a third party.

Dual to this distinction is one drawn by Francis Bacon:

> For although in the more gross sort of fallacies it happeneth, as Seneca maketh the comparison well, as in juggling feats, which, though we know not how they are done, yet we know well it is not as it seemeth to be; yet the more subtle sort of them doth not only put a man beside his answer, but doth many times abuse his judgment.   (Bacon [1915]: 131)

We may summarize this as a distinction between the GROSS fallacy, in which something seems wrong (and is) and the SUBTLE fallacy, in which everything seems OK (but is not). Just as the distinction between sophism and paralogism is a distinction confined to the speaker's intent, with no necessary reflection in the form of the fallacy, so that between gross and subtle fallacies is restricted to the auditor's understanding. Again, the same fallacy could fulfill either role—the sharper your judgment, the less likely it is to be abused. However, at least on the standard treatment, which Bacon's account explicitly echoes, the auditor's understanding, unlike the speaker's intent, is a part of the definition of fallacy, since it determines whether the fallacy seems valid to the auditor.

Reflection on Bacon's discussion may draw attention to a third case, that of SURPRISE, in which something seems wrong (but is not). This is a common phenomenon in mathematics,[3] and perhaps as great a source of error as actual fallacy. A popular supposition is that 'when the results of reasoning and mathematics conflict with experience in the real world, there is probably a fallacy of some sort involved' (Bunch [1982]: 2). This is a plausible inference, but it can be profoundly misleading. Once we outstrip the mathematics consonant with our experience of the real world, our everyday intuitions can no longer guide us. Wilfrid Hodges attributes the frequency of attempted refutations of Cantor's diagonal

---

[3] See Havil [2007] for some excellent examples.

argument to such reasoning. 'Until Cantor first proved his theorem ... nothing like its conclusion was in anybody's mind's eye. And even now we accept it because it is proved, not for any other reason' (Hodges [1998]: 3). As this is often the first such proof enthusiastic amateurs encounter, prominent logicians and journal editors such as Hodges keep receiving their attempted refutations.

The distinction between gross and subtle fallacies, and the existence of surprises, demonstrate that it is not only the method of argument which may seem to be one thing while actually being another. The result of the argument is susceptible to similar confusion. Results which conflict with our prior beliefs will seem to be false, irrespective of their actual truth value. Conversely, results which reinforce those beliefs will be much easier to accept, even when they lack sound justification.

1.3 ARGUMENTATION SCHEMES

Many alternative theories of fallacy have been suggested. One influential approach employs ARGUMENTATION SCHEMES: 'forms of argument that model stereotypical patterns of reasoning' (Walton and Reed [2003]: 195). The underlying idea has a venerable history. It can be traced back to Aristotle's *Topics* and its successive reappropriations in Cicero, Boethius, and mediæval logic (Walton, Reed, and Macagno [2008]: 275ff). More recently, the tradition has been revived in informal logic. Several subtly different characterizations of the argumentation scheme (or argument scheme) are in use. These have given rise to independent, but largely overlapping classifications, one of the most extensive to date being that of Manfred Kienpointner, who distinguishes more than one hundred different schemes (ibid.: 309ff). We shall be following this influential treatment, that of Douglas Walton and collaborators, which has been defended in several books and articles since the early 1990s.

Walton's argumentation schemes are presented as schematic arguments which are typically accompanied by CRITICAL QUESTIONS. The critical questions itemize known vulnerabilities in the argument, to which its proposer should be prepared to respond. In principle, they can always be incorporated into the schematic argument as additional premises (ibid.: 17). This move has advantages in formal implementations of argumentation schemes, but at the cost of obscuring the characteristic dialectical context in which the schemes are typically employed. Many of the schematic arguments are special cases of *modus ponens* in which the hypothetical premise lacks the force of a deductive implication. Hence most of Walton's argumentation schemes are presumptive or defeasible. But deductive inferences can also be understood as argumentation schemes. However, there is a great deal of reasoning in mathematical practice for which informal argumentation schemes would be more appropriate.

A crucial question for the argumentation scheme approach is that of normativity: What makes an instance of an argumentation scheme good or bad? The answer must in part depend on the nature of the scheme. If the scheme is deductive, then it is good precisely when it is valid, as we would expect. Defeasible schemes obviously have less normative force. There are several possible ways in which this might be captured. Broadly, Walton's approach is to say that a scheme is persuasive if all of its critical questions (or at least those where the burden of proof is on the proposer of the argument) receive satisfactory answers, and otherwise not. Hence restating defeasible schemes to incorporate the critical questions as additional premises will have the effect of converting them into deductive schemes (Walton and Reed [2003]: 210).

1.4 FROM ARGUMENTATION SCHEMES TO FALLACIES

Fallacies may be understood as argumentation schemes used inappropriately. This cashes out the troublesome concept of 'seeming valid' in the stereotypical character of the schemes. Essentially, they have been chosen as representative of the sort of arguments generally found convincing.

The use of an argumentation scheme can be fallacious in two distinct ways. Firstly, some schemes are invariably bad. They are distinguished from other invalid arguments by their tempting character, presumably a consequence of their similarity to a valid scheme. This is typical of formal fallacies, such as the quantifier shift fallacy. Secondly, argumentation schemes with legitimate instances can be used inappropriately, characteristically when deployed in circumstances that preclude a satisfactory answer to the critical questions. For example, the argumentation scheme for Appeal to Expert Opinion is associated with a list of questions addressing the expertise of the source of the opinion. The traditional *ad verecundiam* fallacy arises when these questions are not properly answered. One substantial attraction of the argumentation scheme approach to fallacy theory is that it rehabilitates the conventional lists of fallacies as defective examples of defeasible but sometimes persuasive argumentation schemes.

Can all fallacies be analyzed in this way? Walton does not claim this. Indeed, he lists several informal fallacies which he says cannot be reduced to the misuse of a specific argumentation scheme:

(1) Equivocation;
(2) Amphiboly;
(3) Accent;
(4) Begging the Question;
(5) Many Questions;
(6) *Ad Baculum*;
(7) *Ignoratio Elenchi* (Irrelevance); and
(8) *Secundum Quid* (Neglecting Qualifications).

(Walton [1995]: 200f)

Ad baculum seems out of place in this list, since it can be characterized as a defective instance of Walton's scheme for Argument from Threat, which is sometimes admissible in negotiation, but never in persuasion.[4] The other fallacies on this list are of two types. Equivocation, amphiboly, accent, and *secundum quid* are problems that can arise in many different argumentation schemes, and so cannot be characterized by any one scheme. Begging the question, many questions, and *ignoratio elenchi* need extended sequences of argumentation to become manifest, hence they are not associated with any single argumentation scheme either.

Nevertheless, all of these fallacies are within the scope of a slightly more broadly drawn argumentation scheme approach. The former kind, equivocation, amphiboly, accent, and *secundum quid*, can certainly be understood as arising from an inappropriately applied scheme, even if

---

[4]Walton does ask whether the fallacy should be drawn more broadly, to include 'scaremongering tactics that do not involve a threat' (Walton [1995]: 41), but does not answer his own question.

there are several possible schemes which it might be. Moreover, since they mostly seem to turn on definitions or verbal classifications which are ambiguous, insufficiently qualified, or question begging, they might be seen as characteristically besetting the Argument from Verbal Classification. (We shall discuss a question-begging definition as an example of this scheme in the next section.) The latter sort, begging the question, many questions, and *ignoratio elenchi*, require us to broaden our scope from the individual scheme. But each fallacy might still be understood as a pathology of inappropriately applied argumentation schemes, even if we are unable to narrow the blame down to a single scheme.

Hence we have a means of restating the standard treatment without appeal to subjective or psychologistic content. 'Seems good' may be analysed as 'employs argumentation schemes.' We may thereby define a fallacy as a defective instance of argumentation scheme(s) which may or may not be invariably defective.

Since we now have defensible accounts both of how propositions may seem true and of how arguments may seem sound we are in a position to update the typology offered in Table 1. These qualifications are incorporated in Table 2. In this table the combination of a sound method with a false result is labeled ∅, since we now have a rich enough classification to explicitly rule such cases out. If we concentrate on cases of mathematical error which arise where the result is false, whether or not it seems so, a simpler picture emerges, as displayed in Table 3.

1.5 APPLICABILITY TO MATHEMATICS

There are two approaches to the application of argumentation schemes to mathematics. Firstly, we could develop specifically mathematical argumentation schemes which captured patterns of reasoning unlikely to arise in any other domain. Secondly, we could explore the topic-neutral aspects of mathematical reasoning by attempting to apply topic-neutral argumentation schemes. We shall concentrate on the second approach, which promises to help make explicit the connexions between mathematical practice and ordinary reasoning. However, the first approach is also of considerable interest.

Specifically mathematical argumentation schemes have many antecedents, both in argumentation theory and in mathematics. Finely

Method		Result		Outcome
Seems	Is	Seems	Is	
G	G	T	T	Proof
G	G	T	F	∅
G	G	F	T	Surprise
G	G	F	F	∅
G	B	T	T	Howler
G	B	T	F	Subtle Fallacy
G	B	F	T	Howler
G	B	F	F	Gross Fallacy
B	G	T	T	Surprise
B	G	T	F	∅
B	G	F	T	Surprise
B	G	F	F	∅
B	B	T	T	Howler
B	B	T	F	(Tempting) Mistake
B	B	F	T	Howler
B	B	F	F	Mistake

Table 2: A richer typology of mathematical error

tuned argumentation schemes have been developed by proponents of agent-based reasoning in the implementation of expert systems designed to tackle very specific problems. For example, bespoke argumentation schemes have been used to model decision making in the area of organ donation and transplantation (Tolchinsky et al. [2006]). In mathematics, several independent approaches have led to the study of a variety of structures loosely comparable to argumentation schemes. In particular, the automated theorem-proving community has developed 'proof plans' and other models of common fragments of mathematical reasoning (Bundy [1991]). In philosophy of mathematics, the 'inference packages' introduced by Jody Azzouni ([2005]: 9) as 'psychologically-bundled

Method ↘ Result →	Seems True	Seems False
Seems Sound	Subtle Fallacy	Gross Fallacy
Seems Unsound	(Tempting) Mistake	Mistake

Table 3: Mathematical error where result is false

ways of phenomenologically exploring the effect of several assumptions at once' might also be construed as specifically mathematical argumentation schemes.

Many mathematical fallacies exemplify invariably bad argumentation schemes. This is true whether we look for topic-neutral schemes, such as quantifier shift, or more specifically mathematical operations, such as dividing by zero. Are there any mathematical fallacies in which the argumentation scheme is not invariably bad? An historical example of a fallacy manifesting as an inappropriate deployment of a specifically mathematical argumentation scheme may be found in Isaac Newton's argument that the moment of a quantity 'is equal to the moments of each of the generating sides drawn into the indices of the powers of those sides, and into their coefficients continually':

> Any rectangle as AB augmented by a perpetual flux, when, as yet, there wanted of the sides A and B half their moments $\frac{1}{2}a$ and $\frac{1}{2}b$, was $A - \frac{1}{2}a$ into $B - \frac{1}{2}b$, or $AB - \frac{1}{2}aB - \frac{1}{2}bA + \frac{1}{4}ab$; but as soon as the sides A and B are augmented by the other half moments; the rectangle becomes $A + \frac{1}{2}a$ into $B + \frac{1}{2}b$, or $AB + \frac{1}{2}aB + \frac{1}{2}bA + \frac{1}{4}ab$; From this rectangle subduct the former rectangle, and there will remain the excess $aB + bA$. Therefore with the whole increments a and b of the sides, the increment $aB + bA$ of the rectangle is generated. Q.E.D.
> (Newton [1729]: Book II, Lemma II)

This is debunked by George Berkeley as follows:

> But it is plain that the direct and true method to obtain the moment or increment of the rectangle AB, is to take the sides as increased by their whole increments, and so multiply them

> together $A+a$ by $B+b$, the product whereof $AB+aB+bA+ab$ is the augmented rectangle; whence, if we subduct AB the remainder $aB + bA + ab$ will be the true increment of the rectangle, exceeding that which was obtained by the former illegitimate and indirect method by the quantity ab.
>
> (Berkeley [1996]: §9)

Newton employs an ingenious procedure, for which there may be plausible applications, but as Berkeley correctly observes, this is not one of them.

More generally, it is a familiar phenomenon in mathematics that methods of widespread usefulness can produce paradoxical results in a minority of cases (see, for example, Maxwell [1959]: 51, or Barbeau [2000]: 109f). We shall return to this issue in the discussion of howlers. But first we will seek to illustrate by example the topic-neutral approach to the application of argumentation schemes to mathematics.

2 Examples

There is no consensus on how best to classify argumentation schemes. Part of the problem is that there are several mutually independent dimensions of similarity which we might hope that a classification should respect. However, schemes may be loosely grouped in terms of the nature of the conclusions they establish. These include particular and general propositions to be accepted or rejected, actions to be performed, assessments of other arguments, causal claims, rules to be followed or ignored, and commitments to be ascribed to agents. Arguments of most or all of these kinds may be found in mathematical reasoning. Practicing mathematicians have long observed with J.J. Sylvester 'how much observation, divination, induction, experimental trial, and verification, causation, too . . . have to do with the work of the mathematician' (Sylvester [1956]: 1762). Mathematics is not just the derivation of conclusions from axioms.

In this section we shall consider how some specific argumentation schemes may be applied to mathematics. Inevitably, the selection is narrow and unrepresentative, but should give a flavour of the approach.[5]

---

[5] Some further examples have been considered elsewhere, including Appeal to Expert Opinion (Aberdein [2007]: 3ff) and Argument from Sign (Dove [2009]: 143).

2.1 ARGUMENT FROM VERBAL CLASSIFICATION

> *Individual Premise* a has property F.
>
> *Classification Premise* For all x, if x has property F, then x can be classified as having property G.
>
> *Conclusion* a has property G.
>
> CRITICAL QUESTIONS:
>
> (1) What evidence is there that a definitely has property F, as opposed to evidence indicating room for doubt on whether it should be so classified?
>
> (2) Is the verbal classification in the classification premise based merely on an assumption about word usage that is subject to doubt?

Scheme 1: Argumentation scheme for Argument from Verbal Classification (Walton, Reed, and Macagno [2008]: 319)

In many mathematical cases, the premises of the scheme for Argument from Verbal Classification (Scheme 1) will be provably true, and thus the argument deductively sound. But in informal mathematical reasoning this need not be the case. The classification premise might be a hypothesis, well corroborated, but as yet unproven, or a hunch, or even something known to have exceptions, but which seems plausible enough in context. The classification premise could also be a definition. This is harmless if the definition is uncontested, but can give rise to abuse. In Dana Angluin's widely circulated list of spurious proof types, this is neatly characterized as 'proof by semantic shift: Some of the standard but inconvenient definitions are changed for the statement of the result' (Angluin [1983]: 17). Such stipulative definition clearly precludes a satisfactory answer to the second critical question.

On the other hand, many mathematical fallacies turn on question-begging definitions, which smuggle in improbable conclusions disguised as innocuous classification premises. For example, consider Maxwell's

To prove that every point inside a circle lies on its circumference.

GIVEN: A circle of centre O and radius r, and an arbitrary point P inside it.

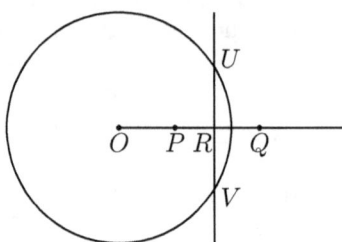

REQUIRED: To prove that P lies on the circumference.

CONSTRUCTION: Let Q be the point on OP produced beyond P such that $OP \cdot OQ = r^2$ and let the perpendicular bisector of PQ cut the circle at U, V. Denote by R the middle point of PQ.

PROOF: $OP = OR - RP$

$OQ = OR + RQ$

$ = OR + RP$ ($RQ = RP$, construction)

$\therefore OP \cdot OQ = (OR - RP)(OR + RP)$

$ = OR^2 - RP^2$

$ = (OU^2 - RU^2) - (PU^2 - RU^2)$ (Pythagoras)

$ = OU^2 - PU^2$

$ = OP \cdot OQ - PU^2$ ($OP \cdot OQ = r^2 = OU^2$)

$\therefore PU = 0$

$\therefore P$ is at U, on the circumference

Example 1: The Fallacy of the Empty Circle (Maxwell [1959]: 18f; see also Bradis, Minkovskii, and Kharcheva [1999]: 25)

'Fallacy of the Empty Circle' (Example 1). This can be translated into the scheme for Argument from Verbal Classification as follows: a is the arbitrary point P; F is the property that there exists a point Q on OP produced beyond P such that $OP \cdot OQ = r^2$; G is the property of being on the circumference. The fallacy may now be clearly understood. The classification premise is true and supported by the proof. But the individual premise is false: F is not a property of arbitrary points, but only of points on the circumference. Supposing otherwise is the critical loss of generality on which the argument turns. The first critical question will clearly receive an unsatisfactory answer.

2.2 ARGUMENT FROM POPULAR OPINION

> *General Acceptance Premise* A is generally accepted as true.
>
> *Presumption Premise* If A is generally accepted as true, that gives a reason in favor of A.
>
> *Conclusion* There is a reason in favor of A.
>
> CRITICAL QUESTIONS:
>
> (1) What evidence, like a poll or an appeal to common knowledge, supports the claim that A is generally accepted as true?
>
> (2) Even if A is generally accepted as true, are there any good reasons for doubting that it is true?

Scheme 2: Argumentation scheme for Argument from Popular Opinion (Walton, Reed, and Macagno [2008]: 311)

There is an important distinction to be made between cases of Argument from Popular Opinion (Scheme 2) where the opinion of the majority is *constitutive* of the truth of A, and all other cases. Thus, when a plebiscite was held to resolve the Schleswig–Holstein question, the opinion of the populace as to whether they were German or Danish was constitutive of the right answer. But a referendum over the moral status of abortion, or indeed the value of π, cannot be defended in this fashion. The popularity of the answer would be independent of its truth, a failure to satisfactorily answer the second critical question.

Are all mathematical questions of the latter sort? No. Exceptions include widely endorsed arbitrary conventions. Many of these, such as axioms, have a claim to self-evidence (whatever that means) that militates against their arbitrariness—they're not true because everyone says so, they're true because they can't but be true. So, not only do we not have reason to doubt them, we have better reasons to believe them than can be provided by this scheme. But there are cases where this scheme may be as good as it gets. For example, the definition of 'straight edge and compass construction.' The impossibility of certain constructions, such as trisecting an arbitrary angle, is a well-established result. But angle trisectors still claim to have produced such constructions, sometimes through sheer incompetence, but in the more adroit cases through equivocation on 'straight edge and compass construction.' They produce a 'construction,' using a 'straight edge' and a 'compass,' but in a non-standard fashion, hence it is not a 'straight edge and compass construction'—a point they may find hard to accept. So, to say that an arbitrary angle cannot be trisected invokes a convention on what counts as a construction. If that's given up, then an arbitrary angle *can* be trisected, but that's not the historically interesting problem. Why not? Because it's not what mathematicians generally accept. So the best argument available in defence of the convention will be an instance of this scheme.

Another example is provided by Hodges's response to analogous attempts to refute the diagonal argument:

> Other authors, less coherently, suggested that Cantor had used the *wrong positive integers*. He should have allowed integers which have infinite decimal expansions to the left, like the p-adic integers. To these people I usually sent the comment that they were quite right, the set of real numbers does have the same cardinality as the set of natural numbers in *their* sense of natural numbers; but the phrase 'natural

number' already has a meaning, and that meaning is not theirs. (Hodges [1998]: 4)

Here Hodges's counter-argument is an instance of Argument from Popular Opinion, whereas the anti-diagonalization argument he is criticizing is an instance of Angluin's 'proof by semantic shift,' that is a defective instance of Argument from Verbal Classification.

2.3 ARGUMENT FROM POPULAR PRACTICE

Major Premise A is a popular practice among those who are familiar with what is acceptable or not in regard to A.

Minor Premise If A is a popular practice among those familiar with what is acceptable or not with regard to A, that gives a reason to think that A is acceptable.

Conclusion Therefore, A is acceptable in this case.

CRITICAL QUESTIONS:

(1) What actions or other indications show that a large majority accepts A?

(2) Even if a large majority accepts A as true, what grounds might there be for thinking they are justified in accepting A?

Scheme 3: Argumentation scheme for Argument from Popular Practice (Walton, Reed, and Macagno [2008]: 314)

This close relative of Argument from Popular Opinion, to which similar considerations apply, arises when it is a practice not a proposition that is at issue. As with that scheme, where majority practice is constitutive of correctness, the scheme for Argument from Popular Practice (Scheme 3) may be legitimately invoked against those whose practice deviates from the majority. An important example is the enforcement of contemporary standards of rigour. Consider the dispute over Wu-Yi Hsiang's alleged proof of Kepler's conjecture, which states that the maximum density of a packing of congruent spheres in three dimensions is $\frac{\pi}{\sqrt{18}}$. In his review of the proof, Gábor Fejes Tóth employs Argument from Popular Practice:

> Hsiang might consider this objection 'a dispute about subjective standards of how much detail a properly written mathematical proof has to contain' . . . However, he has to bear in mind that a mathematical proof is a social process: It is only the acceptance by the mathematical community which affirms the legitimacy of a proof. (Fejes Tóth [1995])

The remark which Fejes Tóth suggests to Hsiang may be understood as an attempt to find a basis for challenging the argument along the lines indicated by the second critical question. If the standards of proof are ultimately subjective, or at least if Hsiang's proof is within the range of subjectively acceptable proof, then the views of the majority are irrelevant. However, as the widespread rejection of Hsiang's proof indicates, this challenge did not succeed. Fejes Tóth is also right about the broader point—mathematical rigour is a popular practice.

In other circumstances, Argument from Popular Practice is notably weak, although it can still be useful, especially in resource-limited contexts. Following the crowd can be the simplest way to get where you want to go. For example, it might be used in defence of time-honoured 'tradecraft': practical heuristics that have been found useful by generations of teachers, students, and professionals, such as the LIATE or ILATE rule for integration by parts. Of course, this scheme can only provide a weak, and potentially misleading justification for such material. Where the heuristic is genuinely useful, there must be a more robust rationale for it. In some cases such practices can degenerate into ritual, precisely because they are warranted *only* by this scheme, the use of which has obscured the absence of proper justification. Such behaviour would exhibit what some mathematics educationalists have referred to as a 'ritual proof scheme.' A proof scheme is not a scheme for any particular proof, but a way of conceptualizing the concept of proof in general. In a ritual proof scheme 'doubts are removed by . . . the ritual of the argument presentation' (Harel and Sowder [1998]: 245).

One might dispute whether Argument from Popular Opinion and Argument from Popular Practice are truly distinct. Indeed, it should be possible to restate each of them in terms of the other. However, at least as far as mathematical practice is concerned, there is useful work for the distinction to do. It is, for example, the distinction between arguing over which practice 'straight edge and compass construction' should refer to, and which of those practices we should follow.

3 Howlers

Having addressed some of the questions raised by fallacies, we shall move on to howlers. These are less immediately problematic, since there is no contradiction in an unsound method leading to a correct conclusion. However, we shall see that they lead to some fundamental questions about mathematical knowledge.

3.1 A SHOPPING LIST AND OTHER HOWLERS

Maxwell introduces his discussion of the howler with the now rather nostalgic example in Example 2. The ingenious schoolboy has discovered a technique much simpler than his teacher's, and which still gets him the right answer. Unfortunately, it does so because of a peculiarity of the formation of the question. Had the shopping list been slightly different then his technique would have been unsuccessful.

Example 3 shows a case in which the implied technique holds greater interest. As Maxwell observes, the perpetrator of this howler has innocently stumbled on something fascinating. Although the technique obviously does not work in general, every quadratic equation can be expressed in a form for which this technique would work: $(1 + q - x)(1 - p + x) = 1 - p + q$, for an equation with roots p, q.

Not all of Maxwell's howlers are of this sort. Some of those 'perpetrated innocently in the course of class study or of examination' (Maxwell [1959]: 88) defy the reconstruction of any underlying technique. In Example 4 one might suspect that the student, having fortuitously struck upon the right answer, has constructed a sequence of *non sequiturs* which superficially resembles proofs he has been taught. Such rote adherence to the irrelevant features of a practice would be another example of the 'ritual proof scheme.'

> *To make out a bill:*
>
> | | |
> |---|---|
> | $\frac{1}{4}$ lb. butter | @ $2s.\ 10d.$ per lb. |
> | $2\frac{1}{2}$ lb. lard | @ $10d.$ per lb. |
> | 3 lb. sugar | @ $3\frac{1}{4}d.$ per lb. |
> | 6 boxes matches | @ $7d.$ per dozen. |
> | 4 packets soap-flakes | @ $2\frac{1}{2}d.$ per packet. |
>
> The solution is
>
> $$8\tfrac{1}{2}d. + 2s.\ 1d. + 9\tfrac{3}{4}d. + 3\tfrac{1}{4}d. + 10d. = 4s.\ 8\tfrac{3}{4}d.$$
>
> One boy, however, avoided the detailed calculations and simply added all the prices on the right:
>
> $$2s.\ 10d. + 10d. + 3\tfrac{1}{4}d. + 7d. + 2\tfrac{1}{2}d. = 4s.\ 8\tfrac{3}{4}d.$$

Example 2: A shopping list—in shillings and pence (Maxwell [1959]: 9)

> *To solve the equation*
>
> $$(x+3)(2-x) = 4.$$
>
> Either $\quad x + 3 = 4 \quad \therefore \quad x = 1,$
> Or $\quad 2 - x = 4 \quad \therefore \quad x = -2.$
>
> Correct.

Example 3: A quadratic howler (Maxwell [1959]: 88)

> *To solve the equation*
>
> $$x^2 + (x+4)^2 = (x+36)^2.$$
>
> [The student proceeds as follows:]
>
> $$x^2 + x^2 + 4^2 = x^2 + 36^2$$
> $$\therefore x^2 + x^2 + 16 = x^2 + 336$$
> $$\therefore x^2 + x^2 - x^2 = 336 - 16$$
> $$\therefore x^4 = 320$$
> $$\therefore x = 80.$$
>
> Correct.

Example 4: A howler defying reconstruction (Maxwell [1959]: 93)

Clearly there are several distinct phenomena that meet our original characterization of the howler, that is that lead by an unsound method to a correct result. Some of these are more epistemologically problematic than others. Firstly, there are cases, such as that discussed last, where there is no discernible method, or if there is a method, it is inherently spurious. This is a straightforward instance of unjustified true belief, which must fall short of knowledge on any conventional definition. Secondly, there are cases where an explicit procedure is followed, which happens to work in the specific case, but not generally. These pose somewhat more of a problem, since, although the more blatant examples, such as those discussed above, might seem to be unjustified, this becomes harder to maintain the more correct cases there are. We might insist that unreliable procedures fall short of the standards of justification required for rigorous mathematical knowledge, but does that mean that the result might still be known in some other, non-rigorous way? Moreover, these cases differ only in degree from procedures that work in all *except* a few specific cases. Provided that the case in question was not one of these exceptions the result would be correct, but would it still qualify as knowledge if the exceptions are not ruled out?

Applied mathematicians are frequently insouciant about such risks:

> [M]any books dealing with Fourier's series continually repeat the condition that the function must not have an infinite number of maxima and minima. We have generally omitted specifying this condition, since no practical function ever does behave in such a manner. Such behaviour is exclusively confined to functions invented by mathematicians for the sake of causing trouble.
> (Eagle [1925], quoted in Barbeau [2000]: 142)

Lastly, there are some procedures which always work, yet we cannot explain why. This was Berkeley's complaint about the calculus. 'For science it cannot be called, when you proceed blindfold, and arrive at the truth not knowing how or by what means' (Berkeley [1996]: §22). By Berkeley's reckoning even these cases would fail to qualify as knowledge, but, despite the manifest shortcomings of Isaac Newton's method, it seems perverse to insist that he did not know that the derivative of x^2 is $2x$. Most failures of rigour are less extreme than that of seventeenth century calculus, but thereby that much more difficult to see as precluding knowledge.

3.2 MATHEMATICAL LUCK

One way of posing the questions raised at the end of the last section is in terms of luck. Is all non-rigorous mathematics lucky? This would include not only historical proofs that fall short of modern standards of rigour, but also contemporary informal mathematical reasoning, unless a rigorized version is readily available. Much of both sorts of mathematics would generally be regarded as suasive, despite not exhibiting maximal rigour, so it seems strange to say its results are obtained by luck. There are several possible answers to this question.

We could just accept the characterization. Perhaps, at least in the strictest sense, such results are lucky. If rigour is required for certainty, and the argumentation in question is non-rigorous, its conclusions cannot be certain. Without certainty, there is always the possibility, however slight, of contradiction. That no such contradiction arises might be regarded as luck—the ability of a conjecture to survive repeated testing does not have the force in mathematics that it has in empirical science.

Long-standing conjectures have turned out not only to have counterexamples, but to have no actual examples. Gian Francesco Malfatti conjectured in 1802 that to maximize the sum of the areas of three circles inscribed in a triangle, each circle must touch exactly two sides of the triangle and both of the other circles. This was accepted for well over a century, until a counterexample was found in the 1960s, and in 1967 it was shown that an arrangement of greater area than any complying with the Malfatti conjecture could be found for any triangle (Aste and Weaire [2000]: 126). Moreover, as we saw in §2.3 above, what counts as rigour is at least in part socially determined and historically contingent. As Roy Sorensen observes, a mathematician developing a novel technique, as Cantor was with his diagonal argument, cannot always anticipate whether it will be accepted by the community (Sorensen [1998]: 332). This makes the cogency of his proof partly a matter of luck.

Alternatively, the characterization could be resisted by a thoroughgoing platonism. The platonist may argue that the proof on the page is not the *real* proof, an abstract object of which it is a more or less imperfect reflection. If the mathematician has genuinely apprehended such an entity, then he can be certain of its results. That the proof he writes down fails to capture every nuance of its platonic counterpart is a comparatively minor, and in principle remediable, detail. However, this degree of metaphysical extravagance poses its own well-known problems.

Another possibility might be to accept lower standards of rigour, so that informal mathematics would count as rigorous. An example of this strategy might be the 'compensation of errors,' whereby errors made at different stages systematically cancel each other out, which Berkeley offered as an explanation for the success of the calculus. However, Berkeley suggested that this approach might yet be formalized to yield a (genuinely) rigorous foundation for the calculus, a proposal taken seriously by some subsequent mathematicians, notably Lazare Carnot, although with little success (Grattan-Guinness [1980]: 102). If such a formalization were achieved without compromising the accepted standards of rigour, then the problem of luck would no longer arise. But in general, the proposal embodies a fundamental and dangerous confusion. One may advocate the use of informal or defeasible systems of reasoning to capture non-rigorous inference in mathematics without remotely suggesting that such systems might be substituted for deductive logic, or obviate the need for mathematical rigour.

Perhaps the underlying difficulty is that we do not have a clear enough understanding of what epistemic luck means, or why it is inconsistent with knowledge.

3.3 EPISTEMIC LUCK

One recent influential treatment distinguishes between six different varieties of epistemic luck (Pritchard [2005]). As we shall see, at least four of these are consistent with knowledge, but at least one of the others is not. Is all mathematical epistemic luck of one of the benign varieties, and if not, can it be safely isolated from mathematical knowledge?

Duncan Pritchard's definitions require some preliminaries, not least to explain how 'lucky' is to be understood. He offers the following:

(L1) If an event is lucky, then it is an event that occurs in the actual world but which does not occur in a wide class of the nearest possible worlds where the relevant initial conditions for that event are the same as in the actual world.

(L2) If an event is lucky, then it is an event that is significant to the agent concerned (or would be significant, were the agent to be availed of the relevant facts).   (Pritchard [2005]: 128, 132)

The second condition is unproblematic, but the first complicates application to mathematics, in which the propositions are necessary. We shall consider how this might be remedied below. However, three of the benign varieties of luck do not attribute luck to propositions directly. We shall address these first.

Capacity Epistemic Luck

'It is lucky that the agent is capable of knowledge' (Pritchard [2005]: 134)

Capacity epistemic luck arises when the agent is only fortuitously in a fit state to acquire knowledge, as for example when he narrowly avoids a fatal accident. Clearly this does not make the knowledge he acquires any less knowledge, whether or not it is mathematical in nature.

Evidential Epistemic Luck

'It is lucky that the agent acquires the evidence she has in favour of her belief' (Pritchard [2005]: 136)

Evidential epistemic luck turns on the lucky apprehension of the evidence supporting the agent's belief. This is also benign. If the evidence provides sufficient justification, the belief is unproblematically knowledge, no matter how lucky the agent was to acquire it. The precise analogue for evidence in mathematical epistemology is open to dispute, but however it is resolved, evidential epistemic luck would not become pernicious. Serendipitous mathematical discoveries would be lucky in this sense, as biographical accounts of mathematical discovery often relate. Perhaps the discoverer combined essential experience of two recondite and unrelated fields, or made a breakthrough as the result of an improbable chain of coincidences.

Doxastic Epistemic Luck

'It is lucky that the agent believes the proposition' (Pritchard [2005]: 138)

Doxastic epistemic luck arises when the agent might in similar circumstances not have formed the belief, despite having all the same data at his disposal. Again, this is consistent with knowledge. It also describes a very familiar experience in mathematics, perhaps more familiar than in everyday knowledge gathering—a novel insight can occur to a mathematician considering a problem which has frustrated dozens of equally well-informed predecessors. Some cruder sorts of howler, where the agent makes a dramatic inferential leap which takes him to the right answer, but which he is wholly unable to explain, might be mistaken for this phenomenon. The crucial difference is between having a compelling proof but being unable to explain how you came by it, which would be benign doxastic epistemic luck, and just gaining a true belief in a flash of insight, which would be unjustified and therefore not knowledge. Doxastic epistemic luck cannot account for cases of insufficient rigour, where the agent is following a familiar procedure, such that his belief formation process is routine and predictable.

Content Epistemic Luck

'It is lucky that the proposition is true' (Pritchard [2005]: 134)

This is the first of three varieties of epistemic luck in which the luck is attached to the proposition, making the application of (L1) problematic for mathematics. As it stands this constraint is trivial for necessary propositions, since these are by definition true in all possible worlds, and thus in all nearby ones. There are several possible avenues for reinterpreting (L1), but, however this is achieved, content epistemic luck is no obstacle to knowledge. The scenario envisaged here is that of a highly unlikely truth, such as a far-fetched coincidence. Clearly if it is true, its unlikelihood is no obstacle to its truth. A special case of content epistemic luck arises in empirical science—it is the basis of the so-called anthropic principle. One might dispute whether such propositions have a place in mathematics, or whether apparent coincidences are merely indications of our limited understanding. But they are not howlers.

We have now surveyed each of Pritchard's uncontroversially harmless varieties of luck, without adequately characterizing the howler. His classification does not claim to be exhaustive—there could be other varieties of benign epistemic luck which are the best fit for some sorts of howler. However, as we shall see, some of the most problematic howlers are well-represented amongst the pernicious forms of luck.

Veritic Epistemic Luck

'It is a matter of luck that the agent's belief is true' (Pritchard [2005]: 146)

The first of these pernicious forms, veritic epistemic luck, arises where the agent has a true belief, for which he has some justification, but he could have had that justification even if the belief were false. Hence, his belief is only true by luck. Much anxiety over insufficient rigour would seem to be of this sort. The concern is that, despite the agent's best efforts, his beliefs may not be true. For example, the 1993 version of Andrew Wiles's proof of the Taniyama–Shimura conjecture for semistable elliptic curves, and thereby of Fermat's Last Theorem, turned out to rest on a 'construction [that] was incomplete and possibly flawed' (Wiles [1995]: 453). Hence his belief in the conjecture was lucky in this sense. Although the conjecture is true, as Wiles would successfully prove the following year, he did not in 1993 have the justification for it that he thought he had. Was his belief unjustified, or merely inadequately rigorous?

Further progress requires that we postpone no longer the reinterpretation of (L1). One option would be to restate the condition in terms of impossible worlds, in which necessary propositions might be false. This would avoid triviality, but generates its own substantial difficulties, including the characterization of 'nearby' in this context. A more attractive approach might be to rewrite the condition without appeal to worlds, possible or otherwise. We might focus instead on variation in the statement of the situation. For example, the shopping list howler would be veritic epistemic luck on this account, since the method would produce the wrong total for most other lists. Similarly the quadratic howler would not correctly factorize equations not of the form $(1 + q - x)(1 - p + x) = 1 - p + q$.

More generally, we want there to be a reliable procedure underpinning the justification of a belief if luck is to be avoided. There are ways of characterizing such a procedure in non-modal terms. For example, we might characterize it as follows. Without too much loss of generality, the derivation of the proposition may be seen as a directed acyclic graph of which the proposition is the unique out-node. (This analysis may be generalized to admit derivations of more complex structure.) For the derivation to be reliable, we may then stipulate that all of the vertices should be true propositions, all of the in-nodes should be uncontroversially justified, and all of the edges should instantiate reliable inferences. That, of course, leads immediately to the question of which inferences are reliable. We might insist on formal derivation, but that would leave no room for informal mathematics. Conversely, we could say that the inferences were reliable if each vertex was in the logical closure of its predecessors. This might suffice for empirical cases, but would be too easily satisfied in mathematics: Goldbach's conjecture may well be within the logical closure of the Peano axioms, but that doesn't mean that we know it. Alternatively, we might make further use of an approach to informal mathematics already discussed in this paper, and identify the reliable inferences as those which correspond to legitimate instances of some agreed types of argumentation scheme. Variations in the level of rigour may then be implemented by varying the set of admissible argumentation schemes, and the threshold of legitimacy for answers to their critical questions. At the strictest level, only constructive inference rules would be admissible, then classically valid rules, and then varying levels of plausible inference reflective of informal mathematical practice. Wiles's 1993 derivation of the restricted Taniyama–Shimura conjecture passed through at least one inference that he was unable to establish to contemporary standards of rigour for mathematical proof, and at least one vertex that may not have been true. Hence, we may see that this derivation would not support a knowledge claim.

Reflective Epistemic Luck

'Given only what the agent is able to know by reflection alone, it is a matter of luck that her belief is true' (Pritchard [2005]: 175)

Reflective epistemic luck is concerned with 'knowledge' obtained without reflective awareness of the process underlying its acquisition. Whether this form of luck is pernicious is a touchstone for the rival internalist and externalist conceptions of knowledge. Internalism requires, as externalism does not, that the knowledge process be reflectively accessible, and hence beliefs which result from this sort of luck would not qualify. The standard example is that of the 'chicken-sexer' who can reliably distinguish male from female chicks, but cannot (accurately) explain how. Externalists accept this as knowledge; internalists demur. Azzouni ([2009]: 18f) considers but rejects an account of informal mathematics as dependent on an analogous subdoxastic recognition of formal derivations. Such an approach would make all informal mathematics exhibit reflective epistemic luck, but presumably without accepting the internalist deprecation of such luck. Yet even from an internalist perspective, there are some putative examples of mathematical knowledge whose contentiousness may be understood in terms of reflective epistemic luck. Beliefs obtained by infallible but inscrutable intuition, and worries about computer-assisted proof are examples.

The sketch of a procedure sufficiently reliable to resist veritic epistemic luck may be adapted to preclude reflective epistemic luck. What would be required is an additional stipulation that the vertices of the graph should be known to the agent by reflection alone. In most examples of informal mathematical practice this condition would not represent much of an additional burden.

If the hope was that mathematical luck would prove to belong exclusively to benign varieties of epistemic luck, then it has been dashed. We have established that epistemic luck represents a genuine problem for mathematical knowledge. However, we have also sketched a possible way out of the problem, which comprises a further application of the argumentation schemes discussed in the first two sections of this paper.

4 Conclusion

In conclusion, these observations on the pathologies of mathematical reasoning have led to several significant theses:

- Sensitive treatment of fallacies and howlers brings to light a richer typology of mathematical error.

- Mathematical fallacies may be better understood in terms of argumentation schemes. This is potentially an extremely powerful device for the understanding of mathematical reasoning, whether formal or informal.

- The howler is an instance of epistemic luck. As such it raises hard questions for the epistemology of mathematics, which argumentation schemes may also help to resolve.

Each thesis opens up a path for future work.

Bibliography

Aberdein, A. [2007]. "Fallacies in mathematics." *Proceedings of the British Society for Research into Learning Mathematics* **27** (3). 1–6.
Angluin, D. [1983]. "How to prove it." *ACM SIGACT News* **15** (1). 16–7.
Aristotle [1995]. "On sophistical refutations." *Fallacies: Classical and Contemporary Readings.* Ed. by H.V. Hansen and R.C. Pinto. Trans. by W.A. Pickard–Cambridge. University Park, PA: Pennsylvania State University Press. 19–38.

Aste, T. and D. Weaire [2000]. *The Pursuit of Perfect Packing*. Bristol: Institute of Physics.

Azzouni, J. [2005]. "Is there still a sense in which mathematics can have foundations?" *Essays on the Foundations of Mathematics and Logic*. Ed. by G. Sica. Monza: Polimetrica International Scientific Publisher. 9–47.

— [2009]. "Why do informal proofs conform to formal norms?" *Foundations of Science* **14** (1–2). 9–26.

Bacon, F. [1915]. *The Advancement of Learning*. First published in 1605. London: J.M. Dent & Sons.

Barbeau, E.J. [2000]. *Mathematical Fallacies, Flaws, and Flimflam*. Washington: Mathematical Association of America.

Berkeley, G. [1996]. "The analyst." *From Kant to Hilbert: A Source Book in the Foundations of Mathematics*. Ed. by W. Ewald. Vol. 1. First published in 1734. Oxford: Oxford University Press. 60–92.

Bradis, V.; V. Minkovskii; and A. Kharcheva, eds. [1999]. *Lapses in Mathematical Reasoning*. First published in 1963. Mineola, NY: Dover.

Bunch, B. [1982]. *Mathematical Fallacies and Paradoxes*. New York, NY: Van Nostrand.

Bundy, A. [1991]. "A science of reasoning." *Computational Logic: Essays in Honor of Alan Robinson*. Ed. by J.-L. Lassez and G. Plotkin. Cambridge, MA: MIT Press. 178–98.

Descartes, R. [1996]. *Meditations on First Philosophy, with Selections from the Objections and Replies*. Ed. and trans. by J. Cottingham. Cambridge: Cambridge University Press.

Dove, I.J. [2009]. "Towards a theory of mathematical argument." *Foundations of Science* **14** (1–2). 137–52.

Eagle, A. [1925]. *A Practical Treatise on Fourier's Theorem and Harmonic Analysis for Physicists and Engineers*. London: Longmans, Green and Co.

Fejes Tóth, G. [1995]. "Review of Hsiang, Wu-Yi, 'On the sphere packing problem and the proof of Kepler's conjecture.' *Internat. J. Math.* 4 (1993), no. 5, 739–831." *Mathematical Reviews* **95g** (52032).

Freud, S. [1990]. *The Question of Lay Analysis*. First published in 1926. New York: W.W. Norton.

Goldstein, R. [1983]. *The Mind Body Problem: A Novel.* New York: Penguin.
Grattan-Guinness, I. [1980]. "The emergence of mathematical analysis and its foundational progress, 1780–1880." *From the Calculus to Set Theory, 1630–1910: An Introductory History.* Ed. by I. Grattan-Guinness. London: Duckworth. 94–148.
Hamblin, C.L. [1970]. *Fallacies.* London: Methuen.
Harel, G. and L. Sowder [1998]. "Students' proof schemes: Results from exploratory studies." *Research on Collegiate Mathematics Education.* Ed. by A.H. Schoenfeld; J. Kaput; and E. Dubinsky. Vol. 3. Providence, RI: AMS/MAA. 234–83.
Havil, J. [2007]. *Nonplussed! Mathematical Proof of Implausible Ideas.* Princeton, NJ: Princeton University Press.
Hodges, W. [1998]. "An editor recalls some hopeless papers." *Bulletin of Symbolic Logic* **4** (1). 1–16.
Holmes, O.W. [1991]. *The Common Law.* First published in 1881. New York: Dover.
Maxwell, E.A. [1959]. *Fallacies in Mathematics.* Cambridge: Cambridge University Press.
Newton, I. [1729]. *The Mathematical Principles of Natural Philosophy [Philosophiæ Naturalis Principia Mathematica].* Trans. by A. Motte. London: Printed for B. Motte.
Pritchard, D. [2005]. *Epistemic Luck.* Oxford: Oxford University Press.
Sorensen, R.A. [1998]. "Logical luck." *Philosophical Quarterly* **48**. 319–34.
Sylvester, J.J. [1956]. "The study that knows nothing of observation." *The World of Mathematics.* Ed. by J.R. Newman. Vol. 3. First published in 1869. New York: Simon & Schuster. 1758–66.
Tolchinsky, P. et al. [2006]. "CBR and argument schemes for collaborative decision making." *Conference on Computational Models of Argument (COMMA 06).* Ed. by P.E Dunne and T.J.M. Bench-Capon. Vol. 144. Frontiers in Artificial Intelligence and Applications. IOS Press. 71–82.
Walton, D.N. [1995]. *A Pragmatic Theory of Fallacy.* Tuscaloosa, AL: University of Alabama Press.

Walton, D.N. and C. Reed [2003]. "Diagramming, argumentation schemes and critical questions." *Anyone Who Has a View: Theoretical Contributions to the Study of Argumentation.* Ed. by F.H. van Eemeren et al. Dordrecht: Kluwer Academic Publishers. 195–211.

Walton, D.N.; C. Reed; and F. Macagno [2008]. *Argumentation Schemes.* Cambridge: Cambridge University Press.

Wiles, A.J. [1995]. "Modular elliptic curves and Fermat's Last Theorem." *Annals of Mathematics* **141**. 443–551.

ESSAY 12

Two Types of Mathematization

Johannes Lenhard
Michael Otte

The quest for certainty is among the most important and fundamental motifs of the modern age. It was a driving factor for the emergence of science and especially shaped the debates about the role mathematics could or should play in science and scientific knowledge. Often mathematics is simply identified with scientific certainty—to the expense of its relevance for the experimental sciences. Alexander underscored this in his statement about the relation between experimental science and mathematics:

> While the experimental philosophers could easily imagine themselves as explorers of the secrets of nature, the case was more difficult for mathematicians. Mathematics, with its

rigorous, formal, and deductive structure, appeared to be an ill-suited terrain for intellectual exploration . . . Mathematicians, it seemed, did not seek out new knowledge or uncover hidden truths in the manner of geographical explorers. Instead, taking Euclidean geometry as their model, they sought to draw true and necessary conclusions from a set of simple assumptions. The strength of mathematics lay in the certainty of its demonstrations and the incontrovertible truth of its claims, not in uncovering new and veiled secrets. (Alexander [2001]: 2)

Contrasting exploration with mathematics, and making that the basic background philosophy might seem rather hasty, however. Especially so as the dichotomy could easily be interpreted the other way around: Mathematics aims at an account of the world in extensive terms and has brought about a huge number of new techniques and experiences—in contrast to Aristotelian science which entered into questions about the true essence of things. And a really revolutionary transformation occurred with respect to the notion of science itself. "In opposition to the logic of . . . the concept of substance, there now appears the logic of the mathematical concept of function" (Cassirer [1953]: 21), which constitutes the model for an explanation of nature in terms of mathematical laws.

Alexander himself acknowledges that mathematics itself changed in the $16^{th}/17^{th}$ centuries and that some mathematicians "joined in the critique of traditional mathematics" and "fashioned themselves as explorers of the unknown" (ibid.: 6). One should perhaps interpret Alexander's observations in a radically different way and consider mathematics from the point of view of mathematization. And there is some truth then in what Alexander has observed, because mathematization did not come in a uniform fashion and we shall try to indicate this comparing the views of Leibniz and Newton, for example.

Mathematics, considered as a human activity, is in itself not homogeneous, but is beset by quite a number of controversies and contrasting views. Consider the following incomplete list of aspects.

First, mathematical certainty and its traditional forms have changed in course of the Scientific Revolution. A new practical mathematics has developed, including the establishment of new connections between mathematical theory and applications—think of analytical geometry and mechanics as two examples among others—the essential point being that mathematics and mechanics have been promoted to the context of philosophy. Descartes, for instance, in legitimating certain methods of construction, which Euclidean mathematics had not admitted, solved quite a number of 'unsolvable' ancient problems. And Mechanics, which had been considered a part of mathematics, became during the $17^{\text{th}}/18^{\text{th}}$ centuries a center around which philosophy should be reorganized. It became pervasive mathematical practice to use mechanical concepts and devices (see Breger [1991]). Serge Moscovici ([1968]) has claimed in fact that the whole scientific revolution consisted in a reorganization of philosophy centered on the new science of mechanics. But it is not quite clear what this means. Does it also include thinking mechanically about the world? For some it did, whereas others were fiercely opposed to such views—think of Leibniz's quarrels with Spinoza on the one hand and with the Newtonians on the other (see Stewart [2006]).

There is also a neo-platonic stream of thought, which tried to unite the teleological and mathematical methods of argumentation, that is, the Aristotelian doctrine of final causes with the mathematical foundation of physics. Leibniz, for example, warned against assigning everything to necessities of matter, retaining final causes even in physics (*Discourse on Metaphysics*: §19). Natural laws, being mere mathematical regularities, Leibniz claims, are contingent themselves and must be justified by substantial reasons. In his fifth reply to Clarke Leibniz says that a law "cannot be regular, without being reasonable; nor natural, unless it can be explained by the natures of creatures" (Ariew [2000]: 64).

Leibniz notion of natural law thus retained some ingredients from the Aristotelian doctrine of causes. A law in order to be "natural" must not only be an "efficient cause", but also a "final cause" in the sense of Aristotle. Spinoza held a similar notion of law of nature (see Miller [2003]). On the other hand Leibniz created the notion of formal proof as we know it today (Hacking [1984]). He thought truth was established by proof and wanted to make communication and evidence completely mechanical, things to which Descartes and Newton were strongly opposed.

Third, the strict separation between the probable and the rational which dominated the Renaissance had also been transformed in the course of the $17^{\text{th}}/18^{\text{th}}$ centuries, bringing the Baconian sciences in the sense of Thomas Kuhn nearer to the classical sciences including mathematics (see for example Hacking [1975]; Daston [1988]; Shapiro [1991]; Shapin [1994]; Serjeantson [1999]; Franklin [2001]; and Serjeantson [2006]). It seems obvious that the notion of scientific or even mathematical objectivity depends on social and technological conditions as much as on reason. Mathematical objects are the outcome of mathematical and cultural practices, not the other way around. The work of Franklin and Serjeantson is particularly illuminating in these respects.

All these observations indicate that the whole process of the formation of mathematized sciences is far from being homogeneous. The differences exemplified by conflicting patterns of mathematization reach back to the quest for individual certainty that marks the Scientific Revolution of the 17th century. From the 16th century on, mathematics became divided into different understandings of mathematical certainty and proof, presenting two options for mathematization:

(O) *Ontological option* Certainty is to be guaranteed by the principles and assumptions (hypotheses) of philosophy and logic.

(M) *Methodological option* Certainty is to be searched for in the constructions and measurements of geometry, mechanics, and empirical application.

Hintikka has on many occasions drawn our attention to the importance of the distinction between "the view of language as the universal medium . . . and the view of language as a calculus" (Hintikka [1997]: 21ff) for which he sees Leibniz as an initial point of reference:

> On the one hand, Leibniz proposed to develop a *characteristica universalis* . . . whose symbolic structure would reflect directly the structure of the world of our concepts. On

the other hand, Leibniz' ambition included the creation of a *calculus ratiocinator* which was conceived of by him as a method of symbolic calculation which would mirror the processes of human reasoning. (Hintikka [1997])

The dichotomic character of Leibniz' project became clearer as soon as it began to be realized in the 19th century and its two components were accordingly taken up by different research traditions, for which the names of Bolzano, and Frege, on the one side, and Boole, Grassmann, Peirce, and Schroeder, on the other side, might serve as first references. And if we look up a logic history book which scholars are mentioned in the book, Bolzano, Frege, Russell, on the one side, or De Morgan, Boole, Jevons, Peirce, and Schroeder, on the other side, depends largely on whether this book had been written by logicians and analytical philosophers or by mathematicians (see for example Kolmogorov and Yushkevich [1992]).

Hintikka has illustrated a closely related dichotomy in terms of the two sides of Leibniz logical project, *characteristica universalis* vs. *calculus ratiocinator*, and our characterization concords with Hintikka. But we do look upon these developments as mathematicians, rather than logicians and we think that the perspective of mathematization helps to better see the continuity of the historical development. What distinguishes modernity since the 19th century from antecedent periods is after all the fact that mathematics itself came to be mathematized and herein people like Bolzano, Grassmann, or Peirce saw the significance of Leibniz for modernity.

Now, from the point of view of mathematization a description in terms of the relationship between individual facts and general laws might perhaps be more illustrative, because it concerns more specifically a historical change from Aristotle's model of demonstrative science to modern theoretical schemata.

To explain individual facts in terms of general principles and intelligible concepts is the primordial concern of one party, whereas to generate general laws from sufficiently rich data structures and factual evidence is the principal interest of the other. This is admittedly a rather superficial first characterization only, given the fact that the concept of a law of nature itself was evolving and was not uniform and was not the same throughout. Let us flesh out the distinction by considering three historical cases from different centuries.

The next section will be devoted to a discussion of Newton and Leibniz, each of which chose and defended one of the options. It shall be argued that both are clearly located in different camps of mathematization. The third chapter will be devoted to Grassmann who can be conceived of as founding father of a structuralistic and formalistic approach which will be analyzed by contrasting it to Ampère's standpoint.

In the course of the second half of the 20th century, things become different, mainly because of the advent of a new instrument of mathematics and science: the computer. With its development, the old complementarity acquires a new form. The fourth chapter will investigate the relation and controversies between Norbert Wiener and John von Neumann. It will reveal that the complementarity between function and structure manifests itself in their conflicting views on modeling strategies, again involving two options for mathematization. Computer technology, it is argued, was a necessary ingredient to revive the two options for mathematization in a distinctively applied context.

1 Leibniz vs. Newton

Any isolated fact allows for infinitely many different explanations. Leibniz emphasized that, no matter what happens or how the data may be, it will always be "possible to find a notion, a rule, or an equation," that is, a law, such that they are not violated. "Thus it can be said that however God might have created the world, it would always have been regular and within some general order. But God chose that world that is . . . simultaneously the simplest in hypotheses and richest in phenomena" (Leibniz [1988]: 44). Thus the law or the theory must be simpler than the reality it is to describe, otherwise it is useless. At least simplicity might be an option to eliminate the arbitrariness of inductive generalization. If any mathematical representation is admissible it is of no objective value, because anything can be described in some way or other. And if an arbitrarily complex theory is permitted then the notion of 'theory' becomes vacuous because there is always a theory. The criterion of simplicity is, however, not without ambiguity and it can easily be used to overrule factual evidence and to make a theory immune to falsification, as in later conventionalism.

Now, the very hallmark of natural philosophy—its commitment to the intelligibility of nature—was radically reinterpreted by and since Newton and mathematics as well as the mathematization of natural phenomena played a fundamental role in this reinterpretation. Mathematics never gives the 'essences' of substances. The foundation of a scientific theory is not to be seen primarily or exclusively in what it describes or explains, that is, in its conformity with observed phenomena and known facts, but is rather to be seen in its fertility and power to make predictions and to discover new facts. This implies that in addition to the procedures that lead to inductive confirmation certain regulative principles are required, like the criterion of simplicity, which was mentioned already, and principles of classification and interpretation.

To Leibniz this was not sufficient, however, to provide real intelligibility. Leibniz could not tolerate a world without reason and human destination, as his troubled discussions with Spinoza and Newton show (see Stewart [2006]: 241). Leibniz abhors a world where all our thoughts and purposes are reduced to mere objects of scientific inquiry. He searched for a thoroughly human and intelligible world, in which even contingent facts would find their explanation, hence his principle of sufficient reason.

Leibniz' determinism is a result of his nominalism and mentalism. In his second letter to Clarke he wrote:

> In order to proceed from mathematics to natural philosophy, another principle is requisite, as I have observed in my *Theodicy:* I mean, the principle of a sufficient reason, viz. that nothing happens without a reason why it should be so, rather than otherwise . . . [B]y that single principle, viz. that there ought to be a sufficient reason why things should be so, and not otherwise, one may demonstrate the being of God, and all the other parts of metaphysics or natural theology; and even, in some measure, those principles of

natural philosophy, that are independent upon mathematics: I mean, the dynamical principles, or the principles of force. (Ariew [2000]: 7)

Nobody could claim, however, that Newton was a man less religious than Leibniz—to the contrary. Newton, Spinoza, and Leibniz were all men with God on their brains. But their notions of God were quite different. What annoyed Leibniz in particular was the more or less positive view of science, which Spinoza as well as Newton derived from their very different views about God (God as Nature itself vs. God as a sovereign and free willed King). Newton conceived of scientific objectivity in terms of lawful explanations, explaining phenomena by inductive generation of mathematical laws. Leibniz considered it a barbarism to attribute forces to material bodies. "To do this was to make something essential to matter that could not be derived intelligibly from its nature" (Brown [1993]: 227). And this was, according to Leibniz, to bring back occult qualities and he consequently attacked Newton's theory of gravitation, resp. the lack of an adequate foundation of it, comparing it to the occult qualities of the Scholastics (Leibniz [1989]: 321ff).

Newton, in contrast to Leibniz, thought that the relationship between mathematics and natural philosophy be methodological rather than ontological (Hacking [1984] and Ihmig [2005]: 247). Newton wanted to end the metaphysics based ideal of science and his *Principia* "marks, conceptually, a radical departure from the then dominant tradition of a mechanical philosophy that explained phenomena, most often qualitatively, by contact forces" (Gingras [2001]: 384f). If the inverse square formula worked, there was no point in speculating about *what gravity really was*. This would be "metaphysics" in the bad old sense. Such was more or less the view of Newton.

With Newton it was to explain nature in mathematical terms, rather than speculating about the essence of things. Thus physics became 'philosophical,' but the new 'natural philosophy' was to be based on observation, mechanical practice and mathematical deduction. Everything that reaches beyond the observable or logical is of a purely hypothetical nature and cannot therefore be considered certain. *Hypotheses non fingo*, Newton famously said.

For whatever is not deduced from the phenomena is to be

> called a hypothesis; and hypotheses, whether metaphysical or physical, whether occult qualities or mechanical, have no place in experimental philosophy. In this philosophy particular propositions are inferred from the phenomena, and afterwards rendered general by induction . . . [T]o us it is enough that gravity does really exist, and act according to the laws which we have explained, and abundantly serves to account for all the motions of the celestial bodies, and of our sea. (*Mathematical Principles of Natural Philosophy*, Book III: General Scholium)

And in a letter to Oldenburg of 1672 Newton exhibited a kind of 'instrumentalism' or positivism, writing that "the proper method of inquiring after the properties of things is to deduce them from experiments" (quoted after McMullin [1990]: 69), that is, define them in terms of the means and procedures of experimentation and mechanical construction. This brings about more pragmatic or positivistic attitudes with respect to the claims of science to certainty and explanation. It is clear, nevertheless, that some a priori ideas about the very nature of the general causes and interpretations of facts must be employed. These ideas in the case of Newton were based on the certain conviction that Nature was essentially mechanical. This does not imply, however, that the genuine sources of motion are material; the essence of matter is inertia which is a merely 'passive principle' (Newton, *Opticks*: 397).

> The formation of the corpuscular-mechanistic world view required rejecting *in toto* Aristotle's conception of an organismic universe and teleological explanatory schema. Rather than merely revising a previously accepted cosmology, it required laying on nature a completely new theoretical framework of interpretation and network of conceptual implications. While Galileo was primarily responsible for dismantling the Aristotelian–Ptolemaic cosmology, the creation of the corpuscular-mechanistic worldview was the achievement of a number of investigators—although it too owed its final culmination to Newton. (Schlagel [1996])

In this sense Moscovici ([1968]) is right when claiming that the Scientific Revolution essentially consisted in raising mechanics to the level of philosophy. Newton's approach to calculus rests, even more firmly than did Barrow's, on the conception of continua as being generated by motion. Newton's exploitation of the kinematical conception went much deeper than had Barrow's. "Despite popular legend, Newton did not create fluxions to accommodate problems involving motion. To the contrary . . . he began with motion and in that he followed a line of thought rooted in the classical sources but given new vitality by the developments in mechanics" (Mahoney [1998]). Newton made a famous dictum to the effect that geometry is founded on mechanical practice. And he criticized those, who, like Descartes and Leibniz, had confused geometry with algebra or arithmetic (for a discussion of these points see M. Domski [2003]: 1114–24). So, different from Leibniz, Newton did not conceive of the continuity principle in metaphysical terms, but made it a part of a bottom-up strategy.

Leibniz, in contrast to Newton, employed metaphysical hypotheses also in establishing the existence of entities, according to his principle that essence precedes existence. Leibniz calls the 'causes of phenomena' true hypotheses (*Nouv. Ess.*, IV: 12) and makes them the cornerstone of his whole philosophy. Leibniz was anti-mechanistic not least in the sense that he kept the Aristotelian distinction between the *generation* of natural things in contrast to the *fabrication* of artificial ones, and he considered the laws of nature as true generators. According to this view, a law must be an active mediator between phenomena, rather than being something different in kind. This means, when looked upon from a different angle, that the particular was recognized only as of a general kind. Aristotelian science treats individuals not in their own right, but as falling under universals, such that induction is not a problem. "Before the 17th century the modern problem of induction did not exist" (Dear [1995]: 15). When experimental science was invented, mathematics became drawn into different directions, as was indicated above in connection with Alexander's thesis. Leibniz retained some features of the Aristotelian model, while Newton established the experimental approach and tried to fit mathematics into it.

Both Newton as well as Leibniz acknowledged that the solution of that methodological problem, call it the problem of induction or of generalization, requires some metaphysical hypothesis. However, Newton considered all hypotheses as private intuitions, which should not interfere with the final presentation of the results of inquiry. He felt that "disputes on hypotheses are vain; for the empirical laws might be explained by alternative hypotheses and any argument from hypotheses can neither raise nor diminish the certainty of these laws" (Chaudbury [1962]: 345).

Leibniz' thinking consisted of symbolic or logical activity and conceptual determinations, devoted to establish coherence and consistency, but it had to be based on metaphysical principles, rather than on mere mathematical ones. Reality as merely factual and given deserves no interest whatsoever. It must be explained by necessary reasoning. Leibniz claimed that mathematical necessity is merely 'moral' or hypothetical, not metaphysical, and further, that mathematics cannot explain anything.

With respect to mathematics itself, we end up with two quite different ideas, which might roughly be characterized by the terms 'conviction' vs. 'evidence.' One group, from Leibniz through Bolzano to the analytical philosophers of the Vienna Circle and their students of today, believes in an analytical ideal of mathematics ruled by formal proof and logical necessity. And as today they can no longer readily presuppose a Leibnizean rational world, a world of essences, they turned to meanings and semantics and finally had to deny that pure mathematics provides knowledge of anything objective at all. Others, from Descartes, Newton and Kant through Peirce and quite a number of modern mathematicians, value insight higher than proof and believe that evidence rules over rigor. There are whole areas of pure mathematics where such views prevail. For example, one can call to mind probability and combinatorics (Gowers [2000]) or of complexity theory (Chaitin [1998]). It is accepted that even pure mathematics may contain contingent facts without reasons and things without a pre-given meaning.

We have associated Leibniz and Newton each with one option of mathematization. The former with the ontological, or top-down and the latter with the methodological bottom-up option. Admittedly, these two camps present 'ideal types.' Although each of these thinkers is most strongly associated with one 'camp,' they will nevertheless probably exhibit traits of both camps. We maintain that it is a systematically valid and fruitful distinction. This claim will be backed by showing in two further case studies how those two camps persist and can be identified in the history of mathematical science.

2 Grassmann vs. Ampère on Electricity

There have been two rather different views of mathematics in the early 19th century—one aspiring toward an arithmetization of mathematics and the other toward a formal axiomatic and a holistic conception of theory—for which the names of Bolzano, on the one hand, and Grassmann, on the other hand, are somewhat representative. With respect to geometry, for instance, one notices that Bolzano did not participate at all in the transformations that led to the new 'golden age' of geometry in the 19th century (Sebestik [1992]) and which resulted in a completely new understanding of what a mathematical theory could be. Grassmann, in contrast, was one of the most important protagonists of this 'revolution,' creating what is called 'linear algebra' nowadays and being, for example, the first to introduce the idea of n-dimensional space (for arbitrary n).

The following example of mathematization is, in fact, provided by Hermann Grassmann's work in mechanics and geometry and in particular in his "Neue Theorie der Elektrodynamik" published in Poggendorff's *Annalen der Physik und Chemie* in 1845. In this paper he applied his new vector algebra, criticizing among other things Ampère's Law of Electrodynamical Force because Ampère had transferred assumptions valid in the area of gravitational force to a very different area ("auf welchem die Elemente mit bestimmten Richtungen begabt sind") where the elements are vectors rather than points. Despite the fact that Ampère had taken great pains to derive his electro-dynamical force law from the results of experiments and mere empirical observation, he nevertheless found it necessary, as Grassmann points out, to introduce the assumption that the force between two current elements acts along the line connecting their midpoints, in supposed analogy to the mode of action of gravitational forces.

Ampère in his important *Théorie mathématique des phénomènes électro-dynamiques uniquement déduite de l'expérience* of 1823 had in fact written: "Guided by the principles of Newtonian philosophy, I have reduced the phenomenon observed by Oerstedt to forces always acting along the line that joins the two particles between which such forces are exerted" (177-8, own translation). This statement appeared in the context of two other additional facts. First Ampère more or less explicitly accused Oerstedt of introducing unnecessary and unwarranted hypotheses, having been influenced by German Naturphilosophie; and second, that mathematical constructivity has no active role to play in finding the laws of nature; rather mathematics is an auxiliary means for expressing that which has been generated inductively from experience alone. Mathematics is just a language.

Grassmann, in contrast, believed that in every new knowledge there is involved a theoretical and conceptual construction and a hypothetical synthesis. Although Ampère always stressed that in establishing the laws of the new phenomena he "had consulted experience alone," he assumed that the direction of the forces must be "necessarily that of the line joining the material points between which the forces are exerted" (Grassmann [1894b]: 149; own translation). Now Grassmann believed that this assumption could not be true. Grassmann, however, accepts Ampère's view that the *form* of the law, which describes the mutual influence of two electrical currents, is to be the same as the form of Newton's law of gravitational interaction between material points. For Grassmann as well as for Ampère, it was very important that the very same mathematical form could represent electro-dynamical attraction and gravitation.

Grassmann, however, believed that the analogy between the laws of gravitation and of electrodynamics had to be based also on a new interpretation of the mathematical objects and operations in question, taking into account that the entities between which electromagnetic forces act are directed quantities, vectors, not points. (A vector is no more an empirical quantity, but is a class of such quantities. In the same manner, a mathematical function in the sense of Cauchy or Dirichlet is no longer an analytical formula, as it was for Euler and Lagrange, but an equivalence class of such formulas, where the equivalence relation is based on the axiom of extensionality).

Grassmann, referring to his *Lineale Ausdehnungslehre* (theory of extension) of 1844, writes:

> For this purpose [to preserve the form of the law, our insertion], I must, however, quote that concept here which I have introduced in a work recently published, doing so before I had any inkling of the new theory. I have indeed proved there that the line segment connecting them must be seen as the product of two points . . . At the same time, I have shown that the area of a parallelogram must be conceived of as the product of the adjoining sides, taking into account these sides' direction and length.
> (Grassmann [1894b]: 155; own translation)

Grassmann had to generalize the idea of a product from numerical quantities to directed ones, to vectors, giving up commutativity. He defines the idea of a product in a completely new and structural way and by this is able to preserve analogy between gravitation and electromagnetism. Because of the fact that the product is to be formed from one of the current vectors and the magnetic field vector of the other current, rather than from the two current vectors, the direction of the resulting force can be parallel to the line joining the infinitesimal elements of electrical currents, as Ampère had assumed. This holds true, for example, if the currents are parallel to each other. This means that in many cases, for instance, when one takes two electrical currents as lying in the same plane, empirical observation does not help in distinguishing between Grassmann's theory and that of Ampère. Grassmann therefore devotes the last quarter of his paper to describing experiments that should help to make the distinction.

Grassmann, Ohm and others from the group of 'abstracting science,' as Caneva ([1975]: 13) has called them, differed fundamentally from the approach of Newton, Cauchy, or Ampère in that they did neither believe in theoretically unaided experiences and observations nor in mathematics as a mere language to report observational facts. Grassmann's approach in particular is based on a hierarchical structure of knowledge, the levels of which interact in a circular process. The levels are the following:

- a general theory of forms or science of symbols, which represents an abstract model for every mathematical theory;

- a specific mathematical theory, such as his vector calculus, for example; and

- the intended applications of that specific theory or calculus to geometry and physics.

Grassmann thus drew a clear distinction in particular between a geometrical theory, like his 'theory of extension,' on the one hand, and geometrical space as a possible object of application of that theory, on the other (for more details, see Otte [1989]).

The group of German thinkers working on 'abstracting science' also adopted a hypothetic-deductive approach, which sometimes even reversed the order of the relationship between theory and experiment. It is clear to them that theory must precede and orient experimentation and that Ampère's procedure, namely inductively searching for a formula that interpolates the given data, is insufficient. Explanation, from such a point of view, means to show how a phenomenon can be subsumed under an axiom or a law of nature. Here mathematical axioms are understood in analogy to natural laws, which seems strange at first sight, because axioms in the modern sense of the term are mere hypothetical assumptions. On the one hand, these hypotheses are not arbitrary, but resume intuition and previous experiences. On the other hand, the laws of nature have also some hypothetical aspects, rather than being absolutely true and certain. The proof of a scientific theory lies in the future, that is, in its capacity to make predictions. And this notion of scientific theory should later on become universal.

The effect was, however, a separation between the humanities and the exact sciences because neither the axiomatical nor arithmetized mathematics fit anymore into the Aristotelian scheme of explanation.

In summary it must be said that both the analytical and synthetic aspects of science and mathematics gained equally in weight and profile. They appear to be reciprocally dependent on each other and circularly connected. This has been called the complementarist character of mathematics (Otte [2003]).

3 Wiener vs. von Neumann: Computer, Mathematics, Models

During the 20th century, the computer influenced science in a way that often has been characterized as a revolution. One prominent change is the emergence and widespread use of computer modeling and simulation. The computer as a new instrument bestows a new meaning upon mathematization and thus sheds new light on the options of mathematization.

It is argued that current simulation methodology still incorporates the two approaches in mathematization which were discussed in the foregoing chapters, albeit in a modified form. This chapter identifies two early and very influential paradigms of computer-based mathematical modeling. The first conceives simulation as the *numerical solution of equations* and can be associated with John von Neumann. It inherits essential traits from the ontological option, in particular the assumption of a correspondence between mathematics and the structure of the world that can be revealed using computational power. The second position, elaborated most prominently by Norbert Wiener, views simulation as *imitating the behavior* of a complex system with a computer model. Here, the technical construction plays a major role, enabling the use of black box approaches that explicitly abandon any ontological stance. The cooperation between Wiener and von Neumann will be discussed, revealing the conflicts between their conceptions.

The major components of the electronic computer were worked out during the 1940s. The development and implementation of various analogue and digital calculators generally took place under the auspices of war-related military research. The beginnings of computer simulation, conceived here as digital technology, make up a part of this history with the mathematicians Norbert Wiener and John von Neumann playing a major role as influential 'founding fathers.' Both were active in wartime research and the corresponding pressure to produce applications exerted an enormous influence on mathematical theory formulation. Both had the vision that the new computer technology combined with corresponding new approaches to mathematical modeling would lead to an epochal reform. They planned to combine their skills with those of other colleagues in an interdisciplinary group. This was the context for the founding of the Teleological Society that led to the Cybernetics Group, which first met secretly in January 1945 (in Princeton), then officially from 1946 onward in New York, and at a further total of 10 Macy conferences. (For more details see the [1993] and [1980] monographs of Heims.) The present paper concentrates on mathematics and, in particular, on the frictions in the conceptions of mathematizing new fields that came to the fore in the controversy between von Neumann and Wiener.

Wiener worked with the engineer Bigelow, commissioned by the National Defense Research Committee (NDRC), on an Anti-Aircraft-Predictor, a computer-based defense system. This work was based directly on Wiener's mathematical-statistical theories on forecasting and filtering which had been formulated to handle incomplete information and partial ignorance. Statistical pattern recognition provided the mathematical theory for dealing with ignorance that, for Wiener, characterized the epistemological human condition in general. The NDRC project also marked the beginning of Wiener's lasting interest in the links between regular processes and goal-directed behavior. He assigned great philosophical significance to these, as recounted in the programmatic paper "Behavior, Purpose, and Teleology" (Rosenblueth, Wiener, and Bigelow [1943]). Through the amalgamation of the new technologies with his theories, Wiener envisaged the onset of a new epoch, the cybernetic age: "If the seventeenth and early eighteenth centuries are the age of clocks, and the later eighteenth and nineteenth centuries constitute the age of steam engines, the present period of time is the age of communication and control" (Wiener [1948]: 39). Although Wiener viewed himself as the decisive founder, Mindell ([2002]) sees this as a self-created foundation myth based on overestimating his own mathematical work and not paying sufficient attention to the context and importance of engineering.

Wiener developed cybernetics as a science that should examine phenomena and models according to their functionality and behavior, and not according to their material and inner structure, reflecting the fundamental human inability to gain real insight into the world's structure. He distinguished between *open box* and *closed box* (or, as more commonly called today, *black box*) approaches. This terminology was adopted from a communication technology using test procedures to evaluate an instrument according to input–output patterns regardless of the mechanism within. Although these types of boxes differ only gradually, this difference describes a typology of competing types of models: The open boxes, which represent and specify more or less detailed mechanisms versus the black box models that can only be treated in terms of their behavior (i.e., functionally) and imitate behavioral patterns without making any statements about the internal dynamics of the phenomena being modeled.

Wiener considers black box modeling to be necessary for philosophical reasons as they reflect uncertainty and also for pragmatic reasons as they provide applicable technology. Modern information theory, developed independently by Shannon and Wiener, reflects this fundamental opacity and deals with information while completely ignoring any meaning. (Humphreys [2004] also pointed to *epistemic opacity* as one, nonetheless deplorable, feature of simulations.) According to this 'statistical' view, a completely random series of numbers maximizes information compared to any 'lawful' series of numbers of the same length. The computer makes it possible to build a mathematizing strategy out of this black box conception. Thereby the missing harmony between world and mathematical models is, in a sense, carried to the extremes: A theory of control and information that explicitly ignores the constitution of objects is in stark contrast to any position in the 'ontological' camp.

John von Neumann, who has made an original contribution to the axiomatics of quantum theory, was at that time a leading mathematician at the Institute for Advanced Studies in Princeton. He also foresaw the fundamental significance of the new computer technology. He was involved particularly in the development of the architecture for the general purpose computer, now named after him, and also in the invention of new simulation approaches (see, e.g., Galison [1996] on the origins of the Monte Carlo simulation in the Manhattan Project). Put briefly, both protagonists were asking themselves what the new scientific discipline would have to look like in order to be able to advance the computer to a new general-purpose instrument.

Despite all they had in common, von Neumann nonetheless adopted a far more formal, one could perhaps even say more optimistic, stance that viewed the computer as an aid in mathematical theory formulation, and engineering applications more as "the rejuvenating return to the source: the reinjection of more or less directly empirical ideas" ("von Neumann: The Mathematician," cited according to Heims [1980]: 121; cf. also Aspray [1990]). Although von Neumann was aware that mathematical modeling had its limits, he believed that these limits could be discovered through knowledge of mathematics and logic. For example, he admired Gödel's incompleteness theorem as an outcome of superior logic that would grant insight into such limits.

In view of the historical remarks at the beginning of this paper, one could count von Neumann as a member of the Leibnizian camp, trusting in homogeneity of world and human (scientific) knowledge, and in the adequacy of theoretical approaches. According to this camp, the generalizing and mathematizing approach of science was particularly successful—at least in principle and potentially—in attempts to condense the laws that govern the world.

John von Neumann suggested a completely different modeling strategy to simulation. This can be seen in a long letter he wrote to Wiener in 1946, that is, at the time when they were planning their project together. In this letter, which remained unknown until Masani ([1990]) found it in Wiener's estate, von Neumann suggested changing the research agenda of the Cybernetics Group. He criticized the choice of the human central nervous system, because it was too complex an object, rendering progress unlikely. Instead, he suggested viruses as objects of research. In this letter, he tried to accommodate Wiener by saying that the virus was a cybernetic object in Wiener's sense (goal-directed, feedback-controlled behavior). However, involuntarily, the letter documents a deep controversy about what mathematical modeling and mathematization is, or should be conceived of.

> What seems worth emphasizing to me is, however, that after the great positive contribution of Turing—cum—Pitts—and—McCulloch is assimilated, the situation is rather worse than better than before. Indeed, these authors have demonstrated in absolute and hopeless generality, that anything and everything Brouwerian can be done by an appropriate mechanism and specifically by a neural mechanism—and that even one, definite mechanism can be "universal." Inverting the argument: Nothing that we may know or learn about the *functioning* [emphasis added] of the organism can give, without "microscopic," cytological work any clues regarding the further details of the neural mechanism.
>
> (Masani [1990]: 243)

Here, von Neumann is alluding to Turing's findings on computability and the universal Turing machine (which corresponds fairly precisely to the independently developed computer) as well as to the neural mechanism of 'firing' as soon as a critical input threshold is reached which had been analyzed by McCulloch and Pitts (who, by the way, were also members of the Cybernetics Group). Networks of neurons can show universal behavior, that is, generate random patterns.

What is remarkable about this excerpt is that it basically contains a complete rejection of Wiener's program of cybernetics. It is precisely *because* of the performance and adjustability of the simulation model that it is impossible to learn anything about true mechanisms from a successful imitation of behavior patterns ('functioning'). The path from a functional to a structural modeling would be cut off; von Neumann saw that completely correctly and thus proposed that the virus should replace the human nervous system as the object of research.

> Now the less-than-cellular organisms of the virus or bacteriophage type do possess the decisive traits of any living

organism: they are self reproductive and they are able to orient themselves in an unorganized milieu, to move towards food, to appropriate it and to use it. Consequently, a 'true' understanding of these organisms may be the first relevant step forward and possibly the greatest step that may at all be required.

I would, however, put on 'true' understanding the most stringent interpretation possible: that is, understanding the organism in the exacting sense in which one may want to understand a detailed drawing of a machine—i.e., finding out where every individual nut and bolt is located, etc.

(Cited according to Masani [1990]: 245)

To follow the rationality of mathematized applied science implies that the goal is 'true understanding,' and that requires working with structurally isomorphic mechanisms. Von Neumann dropped the functional concept of modeling in favor of the structural one. He almost scorned the versatility of functional modeling whereas the imitation ability was the basis for Wiener's high-flying hopes for cybernetics! In sum, von Neumann wanted the computer to inject more instrumental power into the modeling process in order to reach the 'real' structure in complex situations that had been intractable before. Wiener, on the other side, had a different type of generalization in mind, namely a generality that comes out of the versatility of adaptive model behavior. Wiener did not respond to this letter. Basically, both succeeded in pushing through their research agendas: Wiener pursued cybernetics in his sense. Von Neumann went on to carry out separately the proposal planned originally to be performed under one roof—particular mention should be given here to the solution of hydrodynamic equations. A prime example is meteorology, in which it is easy to trace how the controversy flared up over the conception of simulation modeling.

Right from the start, von Neumann considered meteorology, conceived with a complex hydrodynamic system, to be exceptionally suitable for testing the potentials of computer methods. Originally, he had hoped that simulations could deliver heuristic indications that would then lead to improved methods of mathematical analysis (see Goldstine and Neumann [1946]) that alone would be theoretically convincing and adequate. Back in the 1920s, L.F. Richardson had already formulated the so-called fundamental equations of atmospheric circulation, but their complexity prevented him from producing a numerical weather forecast. With Charney as head of a workgroup at the IAS in Princeton, von Neumann readdressed this project as an application for the computer they were developing.

Von Neumann suggested that the mathematical mechanisms (system of PDEs) should be determined first, i.e. a basis should be specified on which a mechanical explanation can rest. These should then be approximated by a discrete system of finite difference equations, and finally this system should be solved through high-speed computing. It is hardly surprising that Wiener was very critical of this approach. For him, it was negligent to aim at a structure, a kind of scientific hubris. Nevertheless, he thought, complex systems were accessible, if only for a statistical approach. He had also viewed meteorological phenomena as a prime example, but in the sense of a statistical pattern analysis that he did not consider to be a second-rank approach, but an unavoidable one. Therefore, he was opposed to von Neumann's 'solution' approach:

> Recently, under the influence of John von Neumann, there has been an attempt to solve the problem of predicting the weather by treating it as something like an astronomical-orbit problem of great complexity. The idea is to put all the initial data on a super computing machine and, by a use of the laws of motion and the equations of hydrodynamics, to grind out the weather for a considerable time in the future... There are clear signs that the statistical element in meteorology cannot be minimized except at the peril of the entire investigation. (Wiener [1956]: 259)

In meteorology, this triggered a long-lasting controversy between statistical and dynamic approaches (see also Wiener [1954]: 252; cf. Dahan [2001] for a discussion of prediction versus understanding). Although it is precisely in meteorology and climate science that von Neumann's approach seems to have been a great success, closer analysis reveals that this success came about only through a strategy of adding Wiener-like approaches, although not in the direction he had anticipated. This is because the long-term stability of the simulation of atmospheric phenomena was attained only in the 1960s through Arakawa's approach that dropped the ideal of the approximation of a solution and added contrafactual assumptions. These brought the necessary stability, and simulation experiments convincingly confirmed that the simulated behavioral patterns adequately reproduced the real ones (Küppers and Lenhard [2005]).

Of course, the controversy between the two modeling approaches is widespread and not limited to meteorology. Von Neumann himself introduced the virus, because he thought it would be possible to model its mechanisms. On the other side, Wiener was preoccupied (along with McCulloch, Pitts, and others) with human neurophysiology. In what sense can the computer simulate intelligence? Right from the beginning, the artificial intelligence (AI) community was divided on this issue. The behaviorist approach had already been taken by Turing (by the way, very familiar with both Wiener and von Neumann), who conceived his test expressly as an 'imitation game.' This black box approach that takes only input–output behavior as criterion is also known as 'weak' AI. This is contrasted with so-called 'strong' AI that insists on using models that implement the correct mechanisms (cf. Shieber [2004]). Hence the positions of weak versus strong AI point to the options for mathematization which this paper discusses.

The two approaches to mathematization that shaped modern science from the beginning on can be retrieved in the case of computer based modeling, albeit in a modified form. Von Neumann was famous for his pragmatic and often ingenious views on the impact of new computing technology on science. He was deeply convinced that speed and capacity of computation would provide a way to enlarge the realm of tractable mathematics, in particular to win new territory in the more complex fields of applied science. For him, this technological revolution takes place in front of a quite stable background, a fundamental harmony between mathematized science and real world phenomena. Thus he adopted a Leibnizian viewpoint as was introduced at the beginning of the paper. Somewhat ironically, Wiener, who always saw Leibniz as his most important forerunner, is located in the other 'camp' of options for mathematization that rejects an ontological fit between mathematics and the world.

As was mentioned already, Wiener held that man has to face a fundamental opacity and strangeness of the world. This fact, however, can be mitigated by technology and science whose force creates some independency from the world's structure. To state it in other words: Due to the versatility and adaptability to behavioral patterns, a technology of black boxes is possible.

The fundamentally new point is that with computer and simulation technology, the opacity ascribed to the external world now also pervades the realm of mathematical modeling. The inner structure and mechanisms of simulation models cannot be 'understood' in the traditional way. However, the technology—including visualization, simulation experimentation, tuning of models by feedback loops, artificial parameterization—compensates for that by presenting a whole methodology that allows the behavior of models to be fitted to that of real-world phenomena. This twist is of utmost significance, because it opens mathematics to the technical constructions in the sense of engineering sciences.

Here the newer developments of simulation technology come into play. Cellular automata (CA) or neural networks (NN) both exhibit a very universal behavior that emerges from tuning, adjustment of parameters, and the like, influenced—but not derived—from theoretical assumptions about structure. These new types of simulation fully reveal its power (cf. F. Rohrlich [1991] and E.F. Fox Keller [2003] for appraisal of CA as simulation in the full sense, cf. also Wolfram's [2002] ambitious claims about 'A New Kind of Science'). These approaches animate and represent the abstract construction of the universal Turing machine on the level of mathematical modeling thus providing a new basis for generalization, namely a type of generalization that accepts assumptions of structure as important, but mainly heuristic input.

This development does not diminish the problems posed by modeling complex phenomena. Visualization and experimentation, the all-important new features of simulation methodology, allow a degree of flexibility and versatility to imitate behavior that was not anticipated by the early proponents of computational modeling. In particular Wiener was strongly influenced by analogous computation devices. At any rate, technology brings into play new strategies of mathematical generalization. The line of argument has made clear that technology, methodology, and epistemology interact.

This new computer-based mathematical approach to generalization in science resembles very much the older views of mathematization employed in the 18th century. The engineering accounts of mathematization had been outdated by Cauchy and his 'camp,' but if we are right in our analysis, the pendulum has swung back and restituted and older balance between mathematics and technology.

Bibliography

Alexander, A.R. [2001]. "Exploration Mathematics: The Rhetoric of Discovery and the Rise of Infinitesimal Methods." *Configurations* **9** (1). 1–36.

Ampère, A.-M. [1823]. "Théorie mathématique des phénomènes électrodynamiques uniquement déduite de l'expérience." *Mémoires de l'Académie royale des sciences de l'Institut de France*. Vol. VI. Reprinted (facsimile) in 1990. Paris: Firmin Didot. 175–388.

Ariew, R., ed. [2000]. *G.W. Leibniz and Samuel Clarke: Correspondence*. Indianapolis: Hackett Publishing Company.

Aspray, W.F. [1990]. *John von Neumann and the Origins of Modern Computing*. Cambridge, MA: MIT Press.

Belhoste, B. [1991]. *Augustin-Louis Cauchy: A Biography*. Trans. by F. Ragland. Berlin: Springer.

Breger, H. [1991]. "Der mechanistische Denkstil in der Mathematik des 17. Jahrhunderts." *Gottfried Wilhelm Leibniz im philosophischen Diskurs über Geometrie und Erfahrung*. Ed. by H. Hecht. Berlin: Akademie Verlag. 15–46.

Brown, S. [1993]. "Leibniz: Modern, Scholastic, or Renaissance Philosopher?" *The Rise of Modern Philosophy*. Ed. by Tom Sorell. Oxford: Clarendon Press. 213–30.

Caneva, K.L. [1975]. "Conceptual and Generational Change in German Physics: The Case of Electricity 1800-1846." PhD thesis. Ann Arbor, MI: Princeton University.

Cassirer, E. [1953]. *Substance and Function*. New York: Dover.

Chaitin, G.J. [1998]. *The Limits of Mathematics: A Course on Information Theory and the Limits of Formal Reasoning*. Berlin and New York: Springer.

Chaudbury, P.J. [1962]. "Newton and Hypotheses." *Philosophiy and Phenomenological Research*. 344–53.

Dahan, A. [2001]. "History and Epistemology of Models: Meteorology as a Case-Study (1946–63)." *Archive for History of Exact Science* **55**. 395–422.

Daston, L. [1988]. *Classical Probability in the Enlightenment*. Princeton, NJ: Princeton University Press.

Dear, P. [1995]. *Discipline and Experience*. Chicago: University of Chicago Press.

Domski, M. [2003]. "The Constructible and the Intelligible in Newton's Philosophy of Geometry." *Philosophy of Science* **70**. 1114–24.

Fox Keller, E. [2003]. "Models, Simulation, and "Computer Experiments"." *The Philosophy of Scientific Experimentation*. Ed. by H. Radder. Pittsburgh: University of Pittsburgh Press. 198–215.

Franklin, J. [2001]. *The Science of Conjecture*. Baltimore, MD: John Hopkins University Press.

Galison, P. [1996]. "Computer Simulations and the Trading Zone." *The Disunity of Science: Boundaries, Contexts, and Power*. Ed. by P. Galison and D.J. Stump. Stanford, CA: Stanford University Press. 118–57.

Gingras, Y. [2001]. "What did Mathematics do to Physics?" *Historia Scientiae* **XXXIX**. 382–416.

Goldstine, H.H. and J. von Neumann [1946]. "On the principles of large scale computing machines." *John von Neumann: Collected Works*. Ed. by A.H. Taub. Vol. 5 (1961). Oxford: Pergamon Press.

Gowers, T. [2000]. "The two cultures of mathematics." *Mathematics: Frontiers and Perspectives*. Ed. by V.I. Arnold et al. American Mathematical Society.

Grassmann, H. [1894a]. "Die Ausdehnungslehre von 1844 und die geometrische Analyse." *Hermann Grassmanns gesammelte mathematische und physikalische Werke*. Ed. by F. Engel. Vol. 1,1. Leipzig: B.G. Teubner.

— [1894b]. "Neue Theorie der Elektrodynamik." *Hermann Grassmanns gesammelte mathematische und physikalische Werke*. Ed. by F. Engel. Vol. 2,2. Leipzig: B.G. Teubner. 147–60.

Hacking, I. [1975]. *The Emergence of Probability*. Cambridge: Cambridge University Press.

— [1984]. "Leibniz and Descartes: Proof and Eternal Truths." *Philosophy through its past*. Ed. by T. Honderich. Harmondsworth: Penguin. 207–24.

Heims, S.J. [1980]. *John von Neumann and Norbert Wiener: From Mathematics to the Technologies of Life and Death*. Cambridge, MA and London: MIT Press.

— [1993]. *Constructing a Social Science for Postwar America: The Cybernetics Group (1946–1953)*. London: Cambridge University Press.

Hintikka, J. [1997]. *Lingua Universalis vs. Calculus Ratiocinator: An Ultimate Presupposition of Twentieth-Century Philosophy*. Vol. I. Dordrecht Kluwer.

Humphreys, P. [2004]. *Extending Ourselves: Computational Science, Empiricism, and Scientific Method*. New York: Oxford University Press.

Ihmig, K.-N. [2005]. "Newton's Program of Mathematizing Nature." *Activity and Sign*. Ed. by M.H.G. Hoffmann; J. Lenhard; and F. Seeger. New York: Springer. 241–61.

Kolmogorov, A.N. and Yushkevich [1992]. *Mathematics of the 19th Century*. Basel and Birkhäuser.

Küppers, G. and J. Lenhard [2005]. "Validation of Simulation: Patterns in the Natural and Social Sciences." *Fournal for Artificial Societies and Social Simulation* **8** (4).

Leibniz, G.W. [1988]. *Discourse on Metaphysics and Related Writings*. Manchester: Manchester University Press.

— [1989]. "Against Barbaric Physics." *Leibniz: Philosophical Essays*. Ed. by R. Ariew and D. Garber. Indianapolis, IN: Hackett Publishing Company. 312–19.

Mahoney, M.S. [1998]. "The Mathematical Realm of Nature." *Cambridge History of Seventeenth-Century Philosophy.* Ed. by D.E. Garber and M. Ayers. Vol. I. Cambridge: Cambridge University Press. 702–55.

Masani, P.R. [1990]. *Norbert Wiener 1894–1964.* Boston and Berlin: Basel and Birkhäuser.

McMullin, E. [1990]. "Conceptions of Science in the Scientific Revolution." *Reappraisals of the Scientific Revolution.* Ed. by D.C. Lindberg and R.S. Westman. Cambridge: Cambridge University Press. 27–91.

Miller, J. [2003]. "Spinoza and the Concept of a Law of Nature." *History of Philosophy Quarterly* **20**. 257–76.

Mindell, D.A. [2002]. *Between Human and Machine: Feedback, Control, and Computing Before Cybernetics.* Baltimore and London: Johns Hopkins University Press.

Moscovici, S. [1968]. *Essai sur l'histoire humaine de la nature.* Paris: Flammarion.

Newton, I. [1952]. *Opticks: or a treatise of the reflections, refractions, inflections and colours of light.* Based on the 4th edition, London, 1730. New York: Dover.

Otte, M. [1989]. "The Ideas of H. Grassmann in the Context of the Mathematical and Philosophical Tradition since Leibniz." *Historia Mathematica* **16**. 1–35.

— [2003]. "Complementarity, Sets, and Numbers." *Educational Studies in Mathematics* **53**. 203–28.

Rohrlich, F. [1991]. "Computer Simulation in the Physical Sciences." *PSA 1990.* Vol. 2. Philosophy of Science Association. 507–18.

Rosenblueth, A.; N. Wiener; and J. Bigelow [1943]. "Behavior, purpose and teleology." *Philosophy of Science* **10**. 18–24.

Schlagel, R. [1996]. *From Myth to Modern Mind: A Study of the Origins and Growth of Scientific Thought.* New York: Lang Publishing Company.

Sebestik, J. [1992]. *Logique et mathematique chez Bernard Bolzano.* Paris: Vrin.

Serjeantson, R. [1999]. "Testimony and Proof in Early-Modern England." *History and Philosophy of Science* **30** (2). 195–236.

— [2006]. "Proof and Persuasion." *The Cambridge History of Science*. Ed. by K. Park and L. Daston. Vol. III. London: Cambridge University Press.

Shapin, S. [1994]. *The Social History of Truth: Civility and Science in Seventeenth-Century England*. Chicago: University of Chicago Press.

Shapiro, B.J. [1991]. *Probability and Certainty in Seventeenth-Century England: A Study of the Relationship between Natural Science, Religion, History, Law, and Literature*. Princeton and New York: Princeton University Press.

Shieber, S., ed. [2004]. *The Turing Test: Verbal Behavior as the Hallmark of Intelligence*. Cambridge, MA and London: MIT Press.

Stewart, M. [2006]. *The Courtier and the Heretic*. New York: W.W. Norton & Co.

Wiener, N. [1948]. *Cybernetics or Communication and control and the Animal and the Machine*. Cambridge, MA: MIT Press.

— [1954]. "Nonlinear Prediction and Dynamics." *Proceedings of the Third Berkely Symposium on Mathematical Statistics and Probability*. 247–52.

— [1956]. *I Am a Mathematician*. Cambridge, MA and London: MIT Press.

Wolfram, S. [2002]. *A New Kind of Science*. Champaign, IL: Wolfram Media.

ESSAY 13

Numerical Analysis and Its (Invisible?) Role in Mathematical Application

Anthony F. Peressini

The obtaining of numerical results in mathematical applications, the so-called "number crunching" is quite often taken for granted, both in "ordinary life" and even in relatively more sophisticated settings in science and engineering. We blithely expect our calculators and computers to produce numerical answers, flawlessly and unambiguously. Whatever mathematics is lurking behind those calculations is hidden, obscured, invisible. This invisible mathematics is known as numerical analysis.

Numerical analysis is not really invisible. Its relative lack of visibility is due largely to inattention. In contemporary science and mathematics, much of the work of numerical analysis escapes attention because it does its job so well. But its history is both visible and perhaps even prominent in the development of modern mathematics. What is more—if one looks in the right places, it is still playing a developing role in burgeoning areas of science.

In this paper I examine numerical analysis—what precisely it is and why it is important. I begin by presenting a selective conceptual reconstruction of one suggestive line in its historical development. I then expand my focus to a general account of what numerical analysis consists today. I suggest a set of concerns and techniques that are characteristic of numerical analysis. Finally I distinguish three different roles numerical analysis plays in application and illustrate how these distinct roles shed light on its nature and varying visibility.

Numerical analysis studies the means of obtaining numerical results for mathematical expressions: Numerical methods are used ingeniously and indirectly to arrive at the actual numbers implied by a formal solution. While contemporary numerical analysis is inextricably wed to the computer, numerical analysis has its origins in the early 16th century's development of tables of logarithms (Goldstine [1977]), which saw mathematicians working to develop the *means* of generating these tables in increasingly clever and indirect ways in order to avoid cumbersome hand calculations. Logarithms were being used as tools to perform what would have been prohibitively difficult calculations—especially in navigation and calculating interest. While it is Napier and Burgi who typically get the lion share of credit for developing logarithms, it was Henry Briggs (1556–1630) who in extending their work, discovered numerical techniques that persist today.

Practically any use of mathematics beyond counting involves numerical analysis. This mathematics is often hidden because it is embedded in the calculators, processors, and other computing machinery used to generate the numerical solutions.[1] Although the application of numerical analysis is rarely seen, it is essential to prediction, confirmation, and even explanation in science, as well as in any other setting in which the mathematics involved goes beyond arithmetic—including a calcula-

---

[1] The notion of a "numerical solution" is importantly ambiguous. As I will develop below it can indicate either the numerical results obtained from the formal (explicit and exact) solution to a problem or the numerical results obtained to a problem without a formal solution.

tor and its square root function, the calculation of which bear not the slightest resemblance to the hand algorithm taught in grade school.

Numerical analysis involves using mathematical techniques to generate numerical solutions to mathematical expressions. If, for example, you used math to determine that you need to build a wall that is $\sqrt{2}$ meters long, you probably used trigonometry and/or calculus. If you go on to actually find the value of $\sqrt{2}$ to nine decimal places, then you used numerical analysis. The modern theory of numerical analysis includes developing numerical methods, but it also seeks to ground the techniques in theory to establish why they work, how accurately, how generally, and how robustly. It works to develop the means of generating numerical solutions not just to simple expressions like $\sqrt{2}$, but also to problems on the cutting edge of science and engineering.

1 *A Conceptual/Historical Example*

In order to develop a better sense of what numerical analysis is (or better yet what numerical analysis does), let me describe a simple example. Consider how square roots are actually calculated.

Naturally, if we need to know a (non-rational) square root, we go to a calculator, but how do calculators (and/or computers) actually solve the problem? A first guess might be that they do so by imitating the old fashion long-division-like algorithm, but they don't. An historical first stop for square root calculations is Babylonia. The Babylonians (circa 1700 BCE) discovered a method for successively approximating the square root of a number Q that involves only division and averaging. The method goes as follows:

(1) Take an initial guess at $\sqrt{Q}$, call it A.

(2) Compute a better guess, B, using the formula, $B = \frac{A+Q/A}{2}$.

(3) Next iterate your refined guess B through the same formula in place of A to get your 3rd guess, $C = \frac{B+Q/B}{2}$.

(4) Repeat until desired accuracy is obtained.

Iteration	Value
1	1.500000000
2	1.416666667
3	1.414215686
4	1.414213562
5	1.414213562
6	1.414213562

Table 1: Babylonian Method

Table 1 shows six iterations of the method for $\sqrt{2}$ and initial guess = 1. You'll notice that the result converges to nine decimal places by the 5th iteration and that the results match the actual value to nine decimal places by the 4th iteration.

The method of successively approximating a value by repeating the same calculation on the result from the last time is called an iterative method. Typically successive approximations are denoted by subscripts (instead of successive letters of the alphabet) so that the third guess is A_3 instead of C and in general the nth guess A_n, thus the formula can be written as $A_{n+1} = \frac{A_n + 2/A_n}{2}$. With respect to iterative methods important questions arise like the following:

(1) For what values of the initial guess will the approximations converge?

(2) How many iterations are required for a desired accuracy?

(3) For what (kind of) roots will the approximations converge?

These are the questions that gave rise to the practice of numerical analysis.

Before moving on to elaborate on the methods of numerical analysis, let me set the stage a bit more by tracing the development of a few more numerical means of calculating square roots.

Notice that finding the value of $\sqrt{2}$ is equivalent to finding the value of x where the function $f(x) = x^2 - 2$ is zero, i.e., solving $x^2 - 2 = 0$. Thus a

Iteration	Low Approx	High Approx
1	1.0	2.0
2	1.0	1.500000000
3	1.250000000	1.500000000
4	1.375000000	1.500000000
5	1.375000000	1.437500000
6	1.406250000	1.437500000
7	1.406250000	1.421875000
8	1.414062500	1.421875000
9	1.414062500	1.417968750
10	1.414062500	1.416015625
11	1.414062500	1.415039062
12	1.414062500	1.414550781

Table 2: Bisection Algorthm

function evaluation problem becomes the problem of finding the zeros of a related function. This illustrates an important technique in numerical analysis, namely, transforming one numerical problem into another in order to bring more powerful techniques to bear. And indeed, casting the problem in terms of finding zeros opens up a wealth of theory and technique.

One of the simplest approaches to finding a zero is known as the bisection algorithm. It is theoretically based in the Intermediate Value Theorem (due to Bolzano, early 19th century). This method begins with two initial guesses (one high and one low) and then splits the difference by squaring the value in the middle and if it is higher than the number whose square root we are seeking, we then keep the midpoint as the new high value for the next iteration. Alternatively, if the midpoint squared is less than the number whose root we seek, then the midpoint becomes the new low value. This is then repeated until desired accuracy is obtained. Table 2 shows the first twelve iterations. Notice how much more slowly it converges—only three decimal places after twelve iterations, compared to nine decimal places after four iterations with the Babylonian Method. (I will return to why this is the case below.) Despite being slower to converge, the bisection algorithm does have a real advantage: It is perfectly stable and is guaranteed to converge.

As a third method, consider a variation of the bisection algorithm—the *Regula Falsi* Method. This method, rather than *bisecting* the interval each time, splits the interval where a "false line" from the function evaluated at the end points of the interval hits the x-axis. As you can see from Figure 1, this way of choosing the next interval, by using more information about the shape of the function, generates intervals that converge more quickly to the actual answer. (The formula for the split point is $c = \frac{ab+2}{a+b}$.) See Table 3 for the first twelve iterations.

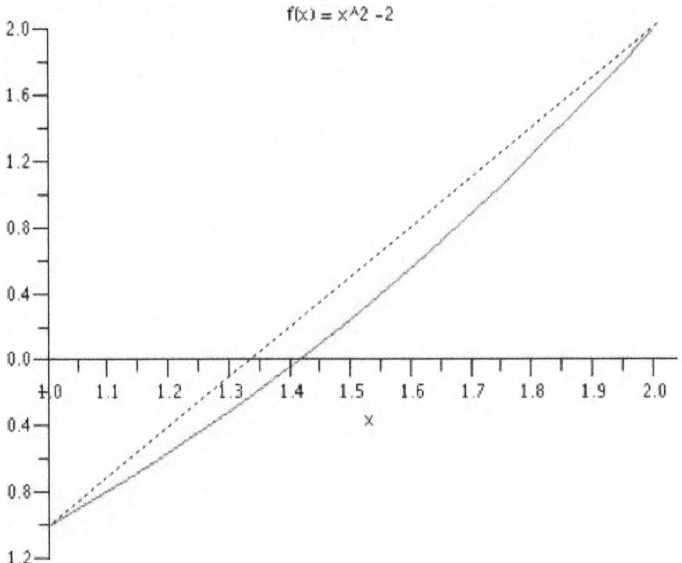

Figure 1: *Regula Falsi* Method

This method is an improvement over the bisection algorithm. While this method doesn't converge qualitatively faster (both methods have what numerical analysts call *linear convergence*), it does converge quan-

Iteration	Low Approx	High Approx
1	1.00000000	2.0
2	1.33333333	2.0
3	1.40000000	2.0
4	1.411764706	2.0
5	1.413793103	2.0
6	1.414141414	2.0
7	1.414201183	2.0
8	1.414211438	2.0
9	1.414213198	2.0
10	1.414213500	2.0
11	1.414213552	2.0
12	1.414213561	2.0

Table 3: *Regula Falsi* Method

titatively faster—by a factor of about four.

An important thing to notice about the *Regula Falsi* method is the way that it refines the general method of the bisection algorithm, taking into account further information and thereby increasing the rate of convergence. This is yet another technique that is typical of numerical analysis.

As a final method of calculating $\sqrt{2}$, let's consider one better than the *Regula Falsi*. While *Regula Falsi* used a false line to narrow the interval, with differential calculus we can actually determine the line tangent to the function at an initial guess and use its x-intercept as the next approximation. As your geometric intuitions might tell you, the x-intercept of such a tangent line closes in on the zero even faster than the false line of *Regula Falsi*. The method is known as Newton–Raphson Method (see Figure 2).

The formula for the Newton–Raphson Method is $A_{n+1} = A_n - \frac{f(A_n)}{f'(A_n)}$. It turns out that the Newton–Raphson method converges *qualitatively* faster than the Bisection or *Regula Falsi* Methods. It has what numer-

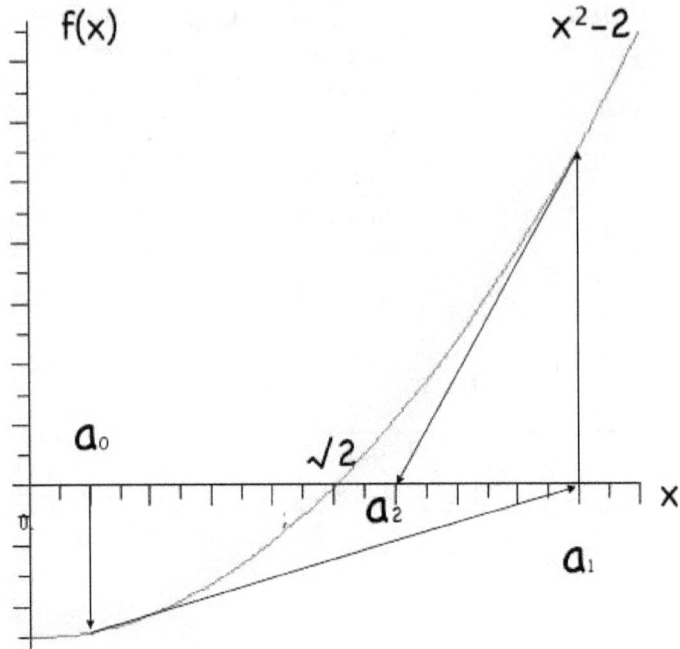

Figure 2: Newton–Raphson Method

ical analysts call *quadratic* convergence for problems like square roots.[2] Notice also that if we substitute the transformed square root function, $x^2 - 2$ in for f, along with its first derivative $2x$, and do a little simplifying, lo and behold, we get the Babylonian Method! It turns out that the Babylonian Method is a special case of Newton–Raphson Method . . . and this explains why it converged so much faster than the bisection or *Regula Falsi* methods. It turns out that the Babylonians discovered a method that it took Newton to generalize by grounding it in the calculus and then only later, modern numerical analysts to explain its rapid

---

[2] Quadratic convergence is essentially an order of magnitude faster than linear convergence. Specifically, if the limit of the ratio of the difference between the converging function and its limit to x^2 is a positive number, then the convergence is quadratic. It is linear if the above obtains with an x instead of x^2.

rate of convergence.

Despite its speedy convergence, the Newton–Raphson Method has a drawback. While numerical analysts have proven that for a sufficiently small interval around a simple zero like $\sqrt{2}$, the method is guaranteed to converge, if an initial guess is not within that critical interval, then strange non-convergent things can happen. Additionally, if the first derivative shares the zero then the method will have additional problems. So the Newton–Raphson Method is somewhat lacking in numerical stability. In actual practice, canned programs for finding a zero use a combination of methods—usually the Newton–Raphson Method and bisection method—for just this reason.

Now that we are back to calculators and zeros—is this then how calculators find square roots? No. We have to go all the way back to one of the people I mentioned above: Henry Briggs and logarithms.

Briggs, in developing his log tables in the 17th century, developed all sorts of ingenious ways to simplify the laborious calculating of huge tables of logarithms. The reason logarithms are useful (and the reason Briggs, Napier, Burgi, and others sought to develop tables of them) is that they simplify calculations by turning multiplication/division into addition/subtraction and exponentiation into multiplication.

The way a calculator comes up with $\sqrt{2}$ is by using the following identities.

$$\sqrt{2} = 2^{1/2} = e^{(1/2) \cdot \log(2)} \qquad (1)$$

$$\log(1+x) = x - \frac{x^2}{2} + \frac{x^3}{3} - \frac{x^4}{4} + \cdots \quad [-1 < x < 1] \qquad (2)$$

$$e^x = 1 + x + \frac{x^2}{2!} + \frac{x^3}{3!} + \frac{x^4}{4!} + \cdots \qquad (3)$$

The calculator calculates $\log(2)$ to machine precision by adding the first few terms of the log series (2). In accordance with (1), this value is then multiplied by 1/2 and the result is exponentiated by adding up the first few terms of the anti-log series (3), again to machine precision.

It is interesting to note that the series approach to logarithms does actually have its origin in Briggs' work. In his ingenious method for computing logarithm tables, Briggs needed to repeatedly calculate square roots and in developing a workable way to do this he discovered the binomial theorem for $(1+x)^{1/2}$, which is the first use of a series calculation of a noninteger exponent. So ironically, Briggs used a series approach in computing his logs, but he used it not to calculate the logs directly, but rather to calculate square roots, which he then used to calculate logarithms.

2 The Field of Numerical Analysis

Having discussed what animates a small corner of numerical analysis, how it developed, and some of its activities, methods, and concepts, I now consider how the field of numerical analysis unfolds into areas beyond calculating square roots.[3]

A standard introduction to numerical analysis will begin with methods of solving equations in one variable—that is, finding the x's, such that $f(x) = 0$. Now as we've seen, the problem of calculating a square root can be transformed into just such a problem, and as you may imagine, so can many others. The methods of solving these equations of one variable get more and more specialized and powerful, and typically the more narrowly one focuses on such a problem, the more efficient the methods get. For example, if one assumes that the function is a polynomial, then the methods available are extremely efficient and stable.

Another cottage industry in numerical analysis is the problem of approximating a function for which one knows only a few of its values. Typically one does this by determining a "well-behaved" function (like a polynomial) that hits all the known data points and then uses this function to estimate unknown data points. This area is known as polynomial approximation and interpolation. Again it generalizes to more general approximation techniques, including using least square methods, orthogonal polynomials, rational functions, and trigonometric polynomials by way of Fourier transformations.

The problems of integration and differentiation are also mainstays of numerical analysis. For any of the real world problems that are solved by calculus (characterized by the operations of differentiation and in-

---

[3] It should be noted that as I move on to contemporary numerical analysis an implicit shift takes place from "pencil and paper" calculations to the electronic computer. While considering this transition is an important one in its own right, I set it aside for the most part in this paper, though I return briefly to it in the final section.

tegration), in order to actually use the solution, numerical results will be needed. Again such methods begin with well-behaved functions of one variable and branch out from there to very complicated functions of multiple variables. In the case of both, the crucial tools are classes of "well-behaved" functions—like the polynomials. It can be proven that for any continuous function of interest, a polynomial can be found that is arbitrarily close to the function of interest at *every* point in a closed interval. So to get a numerical result of a given accuracy to a definite integral or a derivative of a function at a point, one generates an appropriately close polynomial and then treats it as the function of interest to differentiate or integrate. Numerical analysis is further used to prove results concerning error bounds on the difference between the actual and approximate values. Another important area of inquiry concerns how to best choose the original data points by which to generate the approximating function. *Adaptive quadrature* methods chose the location and density of such node points by analyzing the functional variation and using more nodes in regions of higher variation.

One of the most utilized areas in numerical analysis is the solution of systems of linear equations. Equations in multiple unknowns (like on the left below) are cast into matrices (like on the right):

$$x + 3y - 2z = 12$$
$$2x + y - z = 3 \implies \begin{pmatrix} 1 & 3 & -2 & | & 12 \\ 2 & 1 & -1 & | & 3 \\ -1 & -3 & 1 & | & 1 \end{pmatrix}$$
$$-x - 3y + z = 1$$

They are solved by diagonalizing the matrix using row multipliers and row additions to replace existing rows until we have the following diagonal matrix:

$$\begin{pmatrix} 1 & 0 & 0 & | & -2 \\ 0 & 1 & 0 & | & 1 \\ 0 & 0 & 1 & | & 3 \end{pmatrix}$$

This indicates that the solutions to the original equations is $x = -2$, $y = 1$, and $z = 3$. The number of arithmetic operations required by this method (Gaussian Elimination) for a given matrix size is given in Table 4 (Burden, Faires, and Reynolds [1981]: 272).

Matrix size ($n \cdot n$)	Multiplications/ Divisions	Additions/ Subtractions
3	17	11
10	430	375
50	44,150	42,875
100	343,300	338,250

Table 4: Gaussian Elimination Operations

Obviously this ends up requiring a lot of arithmetic—and the problem is that each operation introduces rounding error, which can under certain conditions practically swamp out the solution, rendering it worthless. Methods for solving many other kinds of problems require the solution of very large (hundreds of thousands) systems of equations, hence means of reducing the storage space and operations required are as important as accuracy. Numerical analysts continue to develop more efficient methods requiring less storage and fewer operations. Additionally they have developed means of characterizing the matrix with respect to how it will respond to Gaussian Elimination Methods. Depending on its "condition," supplementary matrix treatments may be used, such as maximal or scaled pivoting to reduce rounding error, which is highly dependent on the relative size of the matrix elements and their position in the matrix.

Important results in numerical matrix theory have also been developed for matrices with special structure (e.g., upper triangular, tridiagonal, Hessenberg, diagonally dominant, positive definite, and sparse). In these cases, theorems can be proven to show that variations of Gaussian elimination are stable and will not require row interchanges because of small or zero elements in key positions. One of the more important developments in numerical linear equations is *iterative techniques*, much like the ones for the square root function considered above. These methods are particularly important for sparse and other specially structured matrices that are generated by methods of solving ordinary and partial differential equations. (See Stewart [1973] for details.)

Differential equations are one of the most ubiquitous mathematical tools in all of applied mathematics. There are ordinary and partial differential equations and among these there are initial value problems and boundary value problems. And then there are systems of each of these kinds as well. These equations arise both in describing the fundamental laws of nature and also in very application-oriented engineering contexts. I won't go any further into these here, but this area is one of the more lively areas in numerical analysis today. Many of the methods of numerically solving differential equations involve solving massive systems of linear equations and/or numerical integration and differentiation. Hence again, we see numerical analysis transforming one kind of problem into another more numerically tractable one.

Having presented the major problems of numerical analysis let me now summarize a bit. Numerical analysis involves problems like approximating a function using "simpler" functions, calculating numerical derivatives, solving initial value and boundary value problems for differential and partial differential equations, finding values for definite integrals, solving systems of linear equations, finding fixed points and characteristic values for functions, and fitting curves to data sets. Numerical techniques are often *iterative*. An iterative technique involves performing a fixed series of arithmetic calculations, called iteration, on an initial approximation. As the iterations progress, the approximation gets "nearer" the actual solution. Theorems are then sought to establish results such as

- the conditions under which the series of iterations will converge for particular initial values;

- restrictions on the initial value;

- the rate at which the series will converge;

- the accuracy of the approximation;

- the sensitivity to round-off error; and

- the average and maximum number of iterations required for a given accuracy.

It is worth pointing out that while my distilled account of the field of numerical analysis might suggest that its major problems have been worked out, such is not the case. In each of the areas I mentioned (and others I did not) work continues on particularly recalcitrant classes of problems. This work largely consists of discovering and characterizing such numerically ill-conditioned problems and developing more subtle, efficient, and effective means of obtaining results.

3 Numerical Analysis as Applied Mathematics

In the times of Briggs, Newton, and Bolzano (to name a few) the relationship between the formal analytic problems and the numerical were still being worked out. In the wonderful stewpot that was mathematics from the 16th century on, we see the concurrent development of formal notation, logical foundations, analysis, and the topology of the real numbers, which are precisely the theoretical underpinnings of contemporary numerical analysis. In this period numerical concerns were naturally quite visible.

I'd like to suggest that some of this persists today. Let me offer the following three rough and ready ways in which numerical analysis functions in scientific application to draw this out.

Instrumental Exact solutions are available, e.g., finding the zeros of a complicated polynomial.

Essential No exact solution is available (even in principle), e.g., certain differential equations.

Explanatory/exploratory The problem is sufficiently complicated that the relationship between the formal mathematical and the numerical is not understood, e.g., general relativity theory.

The first role is instrumental in that a closed exact solution is known and the numerical methods are used only to "do the calculations." Imagine an engineer needing to find the zeros of a complicated cubic polynomial—messy work, but work that could be done without numerical analysis. Things get a bit more fuzzy when one has a simple first-order homogenous differential equation that has a *trigonometric* solution: Obtaining the exact solution requires no numerical analysis, but evaluating the trigonometric solution at values of interest does.

But more often than not, even in well-understood settings, exact solutions are unavailable. In this case the role of numerical analysis is essential.[4] Consider something as straightforward as a second-order differential equation with initial values (a so called "initial value problem"). Such a problem might arise in the modeling of the efficacy of the brakes of a car, taking into account wind resistance and friction. When exact solutions are not available (as they mostly are not), then the only way to

(1) confirm that the differential equation models the braking situation and

(2) make use of the differential equation to determine a breaking distance

is by way of numerical analysis and its highly theoretical grounding in pure mathematics. Nonetheless, the numerical analysis required to perform (1) and (2) is understood to the point that it now covered in "first courses" in numerical analysis. So while essential, these applications of numerical analysis are so routine as to be "invisible."

The third explanatory/exploratory way is important here because numerical analysis applications of this kind are clearly *not invisible* in any sense. These are examples in which the applied mathematical model and its analytic solutions are sufficiently complicated that the relationship between the mathematical model and numerical results are not understood very well at all. Examples like this seem to be present in general relativity, quantum mechanics, and even some classical systems that are chaotic or Brownian.

For example, in the study of the dynamics of deterministic systems using differential equations, the numerical methods are well understood. It turns out, however, that most real phenomena have a stochastic or random component; the mathematical tools used to describe such stochastic dynamics are stochastic differential equations. Many of the

---

[4]I intentionally didn't use the term "indispensable." There are philosophers who think that indispensability is the mark of something metaphysically significant for mathematics. I don't, but for this discussion it doesn't matter. See Peressini [1997] and [2003] for indispensability issues.

numerical methods developed to handle the dynamics of deterministic systems with analytic solutions have been *applied* to systems with stochastic solutions, but relatively little is yet known of the behavior (how stable, how convergent, etc.) of such methods in stochastic settings. In general, whenever numerical methods developed on classical (Newtonian) physics are used outside their theoretically understood domain of application—with theory struggling to catch up—we find numerical analysis playing this explanatory/exploratory role. (For specific examples see Montaldi [2000]; Sprekels [2003]; and Burrage, Lenane, and Lythe [2007].)

The explanatory/exploratory use of numerical analysis is even more pronounced in general relativity. Within the study of general relativity a focus called *numerical relativity* has developed. (For specifics and discussion see Arnold, Ashtekar, and Laguna [2002]; Gentle [2002]; and Carlson [2006].) The study of gravity waves involves looking at massive objects (black holes or neutron stars) in very close proximity, since it is only under such conditions that waves large enough to be detected will be produced. While the formal solutions to such problems are straightforward, obtaining numerical results for such a system is incredibly difficult due to a variety of factors that are not fully understood. The numerical analysis being done in this case is unusual in that it is not being approached by bracketing the problem and then applying the pure analytic theory of numerical analysis. Rather, (i) the science is being reconsidered in the hope of discovering some physically motivated insight or simplification or reduction that will render the numerical problem tractable, and (ii) the application of the (unsuccessful) numerical methods to the actual problem *are being simulated* on related but simpler and better understood problems in order to gain insight into why the method goes fails on the actual problem.

The explanatory/exploratory role of numerical analysis as in the above examples is characterized by the fact that its presence is asserted (if you will) to such an extent that it has its own rich relationship with the two traditional players: the formal applied mathematics and the physical world. In the more familiar roles (instrumental and essential), we see the reciprocal movement between the formal theory and the physical world in which theory is confirmed by its prediction/explanation of physical phenomena and the further enriching, expanding, and revising of the formal theory inspired/necessitated by additions and complications in the physical phenomena, ..., and so on back and forth. But in settings in which numerical analysis is playing an explanatory/exploratory role, the relationship between the formal theory and the physical world is richly mediated by the numerical analysis. In such cases the formal theory's numerical behavior is not fully understood. It requires the explanatory and exploratory machinery of numerical analysis to shed light on it in order for the model to make tractable predictions. Similarly, numerical insights can motivate mathematically equivalent ways of decomposing or understanding the formal model that can in turn shed light on the physical system itself.

It is worth mentioning here that the role of the computer in the development of numerical analysis should not be understated, but neither should it be taken as single-handedly making numerical analysis possible. As we have seen, remarkable developments in numerical analysis happened before the advent of computing machinery and indeed were necessary precisely because certain calculations were cumbersome to the point of being impractical in the absence of computers. So while numerical analysis itself is not essentially tied to computers, nonetheless, computers are responsible for the relatively recent emergence of the explanatory/exploratory role of numerical analysis. In particular, the very possibility of numerically exploring these contemporary models of physical systems takes for granted the availability of computers.

What is going on in the explanatory/exploratory category resembles what was going on in the more rough and tumble days of the 16th and 17th centuries, where the path from mathematical expression to numerical results was largely uncharted. It is informative to consider why it might be that numerical analysis is particularly visible in these two rather distinct settings.

In the period before the rigorous formalization of mathematics, perhaps as far back as Babylonia through the early 17th century (the cut off here is somewhat arbitrary), it is rather anachronistic to interpret any of the work being done as *numerical analysis*. To be sure much of this early work was foundational for numerical analysis but only because it was foundational for the rigorous formal axiomatic mathematics we have today. Prior to the formal development of analysis, while one can point out mathematical work on numerical techniques, properly speaking, no numerical *analysis* was being done since the means of comparing and classifying convergence rates and error analysis were not available. During this period (and ensuing periods leading to modern numerical analysis) the *visibility* of work on numerical methods was pervasive due to its being explored concurrently with the development of analysis in general.

In contrast, contemporary numerical analysis' visibility is limited largely (as we have seen) to its explanatory/exploratory role in scientific application. In the instrumental and even essential roles characterized above, its invisibility is due to the numerical tractability of the theory employing numerical analysis. In such settings, the numerical landscape is charted and it can fade invisibly into the background. In other application settings, ones that are numerically problematic, numerical analysis plays an active and visible role.

This difference in the visibility and role of contemporary numerical work in mathematics grew out of two distinct (though perhaps not independent) developments: the modern formalization of analysis and the advent of the computer. Once in place, these two developments allowed the three roles of numerical analysis described above to coalesce into more or less the form we find them in now.

Acknowledgments

I'd like to thank an anonymous referee for many helpful comments and Conor Peressini for technical assistance.

Bibliography

Arnold, D.; A. Ashtekar; and P. Laguna [2002]. "Proceedings of The Institute for Mathematics and its Applications Workshop on Numerical Relativity." URL: http://www.ima.umn.edu/nr/ (visited July 2007).

Burden, R.; J. Faires; and A. Reynolds [1981]. *Numerical Analysis.* 2nd ed. Boston: Prindle, Weber & Schmidt.

Burrage, K.; I. Lenane; and G. Lythe [2007]. "Numerical methods for second-order stochastic differential equations." *SIAM Journal on Scientific Computing* **29**. 245–64.

Carlson, E. [2006]. "Numerical General Relativity." *Eric Carlson Home Page.* URL: http://www.wfu.edu/~ecarlson/numerical.html (visited July 2007).

Gentle, P. [2002]. "Regge Calculus: A Unique Tool for Numerical Relativity." *General Relativity and Gravitation* **34**. 1701–18.

Goldstine, H. [1977]. *A History of Numerical Analysis from the 16th Century through the 19th Century.* New York: Springer Verlag.

Montaldi, J., ed. [2000]. "Numerical Analysis." *Mechanics and Symmetry in Europe Home Page.* URL: http://www.maths.manchester.ac.uk/~jm/MASIE/projects/sec2.html (visited July 2007).

Peressini, A. [1997]. "Troubles with Indispensability: Applying Pure Mathematics in Physical Theory." *Philosophia Mathematica* **5** (3). 210–27.

— [2003]. "Mark Colyvan's The Indispensability of Mathematics." *Philosophia Mathematica* **11** (3). 208–33.

Sprekels, J., ed. [2003]. "Numberical analysis of complex stochastic models." *Weierstrass Institute for Applied Analysis and Stochastics Home Page.* URL: http://www.wias-berlin.de/publications/annual_reports/2002/node77.html (visited July 2007).

Stewart, G. [1973]. *Introduction to Matrix Computations.* Orlando, FL: Academic Press.

350

About the Authors

ESSAY 1

JOHN DAWSON is Professor Emeritus of Mathematics at the Pennsylvania State University. He is author of *Logical Dilemmas: The Life and Work of Kurt Gödel* and was co-editor of Gödel's *Collected Works* and of the journal *History and Philosophy of Logic*.

jwd7@psu.edu
http://www2.yk.psu.edu/~jwd7/

Penn State York
York, PA, USA

ESSAY 2

MADELINE MUNTERSBJORN is Associate Professor of Philosophy at the University of Toledo. In an effort to understand how knowledge grows, she studies the history of calculus, the origins of natural selection and the social significance of science-fiction. These interests have common origins in the Latin verb *monstrare*, which means "to instruct or make evident," from which both *demonstrations* and *monsters* descend.

Madeline.Muntersbjorn@utoledo.edu   University of Toledo
Toledo, OH, USA

ESSAY 3

IAN DOVE is Associate Professor and Director of Logic and Critical Thinking at the University of Nevada. He received his PhD in 2003 from Rice University for research on the varieties of mathematical fallibilism. His current research includes work on the use of formal methods in informal logic and the philosophical consequences of informal methods in mathematics and the philosophy of applied mathematics.

ian.dove@unlv.edu   Department of Philosophy
University of Nevada
Las Vegas, NV, USA

ESSAY 4

KOEN VERVLOESEM has a master's degree of engineering in computer science, with a specialization in artificial intelligence. He also holds a master's degree of philosophy, with a thesis on the role of computers in proving mathematical theorems.

koen@vervloesem.eu

ESSAY 5

GIANLUIGI OLIVERI received a D.Phil in philosophy at Oxford under the supervision of Michael Dummett and did research in philosophy of mathematics, particularly on the realism/anti-realism debate, and in philosophy of language, including the impact of naturalism on philosophy of language. He is author of *A Realist Philosophy of Mathematics* and co-editor of *Truth in Mathematics*, and *The Philosophy of Michael Dummett*.

gianluigi.oliveri@unipa.it   University of Palermo
Palermo, Italy

ESSAY 6

JOSÉ FERREIRÓS is Professor of Logic and Philosophy of Science at the Universidad de Sevilla. He is the author of *Labyrinth of Thought: A history of set theory and its role in modern mathematics*, and co-editor of *The Architecture of Modern Mathematics: Essays in history and philosophy*.

josef@us.es

Departamento de Filosofia y Logica
Universidad de Sevilla
Sevilla, Spain

ESSAY 7

AHTI-VEIKKO PIETARINEN (PhD, University of Helsinki, 2002) is Professor of Semiotics. He specializes in logic, philosophy of maths, philosophy of language, Peirce, and pragmatism. He has recent publications in *Journal of the History of Ideas*, *Semiotica*, *Perspectives on Science*, *Studia Logica*, *History of Philosophy Quarterly*, and *Pragmatics & Cognition*. His monograph is *Signs of Logic* (Synthese Library, Springer).

ahti-veikko.pietarinen@helsinki.fi

Department of Philosophy
Institute for Art Studies
University of Helsinki
Helsinki, Finland

ESSAY 8

BENEDIKT LÖWE is *Universitair Docent* at the Institute for Logic, Language and Computation of the Universiteit van Amsterdam, Scientific Director of the Graduate Programme in Logic and Professor of Mathematics at the Universität Hamburg.

bloewe@science.uva.nl
http://staff.science.uva.nl/~bloewe/

Institute for Logic, Language and Computation
Universiteit van Amsterdam
Amsterdam, The Netherlands

Department Mathematik
Universität Hamburg
Hamburg, Germany

About the Authors

THOMAS MÜLLER is *Universitair Docent* at the Department of Philosophy of the Universiteit Utrecht and *Privatdozent* for Philosophy at the Universität Bonn.

Thomas.Mueller@phil.uu.nl
http://www.phil.uu.nl/~tmueller/

Departement Wijsbegeerte
Universiteit Utrecht
Utrecht, The Netherlands

Institut für Philosophie
Rheinische Friedrich-Wilhelms-Universität Bonn
Bonn, Germany

EVA MÜLLER-HILL is a PhD student at the Institute for Philosophy at the University of Bonn, and lecturer for Philosophical Logic at the Department of Philosophy at the University of Cologne.

eva.wilhelmus@uni-koeln.de

Institut für Philosophie
Rheinische Friedrich-Wilhelms-Universität Bonn
Bonn, Germany

Philosophisches Seminar
Universität zu Köln
Köln, Germany

ESSAY 9

ROY WAGNER received a mathematics PhD in 1997 and a philosophy PhD ten years later. He publishes papers in mathematics, philosophy of mathematics, and critical theory. Wagner held visiting positions in Paris VI, Cambridge University, Boston University, and the Max Planck Institute for the History of Science. He currently teaches at the school of computer science in the Academic College of Tel-Aviv-Jaffa.

rwagner@mta.ac.il
http://www2.mta.ac.il/~rwagner

The Cohn Institute for the History and Philosophy of Science
Tel Aviv University
Tel Aviv, Israel

ESSAY 10

DANIELLE MACBETH is T. Wistar Brown Professor of Philosophy and author of *Frege's Logic* (Harvard University Press, 2005). She has also published on a range of issues in the philosophy of mathematics, and on other topics.

dmacbeth@haverford.edu
http://www.haverford.edu/phil/
faculty/macbethpage.htm

Haverford College
Haverford, PA, USA

ESSAY 11

ANDREW ABERDEIN is Associate Professor of Logic and Humanities at Florida Institute of Technology. He received his PhD in 2001 from the University of St Andrews for a thesis on the development of nonclassical logics. His current research includes work on the foundations of argumentation theory and the epistemology of mathematics.

aberdein@fit.edu
http://my.fit.edu/~aberdein/

Department of Humanities and Communication
Florida Institute of Technology
Melbourne, FL, USA

ESSAY 12

JOHANNES LENHARD is affiliated with the Philosophy Department of Bielefeld University and research associate professor at the University of South Carolina in 2009–10. Philosophical issues of mathematics are a main research interest for him. He is specially focused on applied mathematics and, most recently, computer simulation.

johannes.lenhard@uni-bielefeld.de
http://www.uni-bielefeld.de/iwt/
personen/lenhard/

Bielefeld University
Bielefeld, Germany

MICHAEL OTTE is Emeritus Professor at the University of Bielefeld. He has published articles in pure mathematics (complex analysis and Lie group theory), mathematical education, epistemology and the history of mathematics. A recent book (*Sign and Activity*, Springer, 2005) honors his 65th birthday.

michaelontra@aol.com   Bielefeld University
    Bielefeld, Germany

ESSAY 13

ANTHONY PERESSINI is Associate Professor in the Philosophy department at Marquette University. His research interests include Philosophy of Math and Science, and Philosophy of Mind/Consciousness.

anthony.peressini@marquette.edu   Marquette University
    Milwaukee, WI, USA

www.ingramcontent.com/pod-product-compliance
Lightning Source LLC
Chambersburg PA
CBHW050330230426
43663CB00010B/1808